Oxford Lecture Series in
Mathematics and its Applications 23

Series editors

John Ball Dominic Welsh

OXFORD LECTURE SERIES IN MATHEMATICS AND ITS APPLICATIONS

1. J. C. Baez (ed.): *Knots and quantum gravity*
2. I. Fonseca and W. Gangbo: *Degree theory in analysis and applications*
3. P.-L. Lions: *Mathematical topics in fluid mechanics, Vol. 1: Incompressible models*
4. J. E. Beasley (ed.): *Advances in linear and integer programming*
5. L. W. Beineke and R. J. Wilson (eds): *Graph connections: Relationships between graph theory and other areas of mathematics*
6. I. Anderson: *Combinatorial designs and tournaments*
7. G. David and S. W. Semmes: *Fractured fractals and broken dreams*
8. Oliver Pretzel: *Codes and algebraic curves*
9. M. Karpinski and W. Rytter: *Fast parallel algorithms for graph matching problems*
10. P.-L. Lions: *Mathematical topics in fluid mechanics, Vol. 2: Compressible models*
11. W. T. Tutte: *Graph theory as I have known it*
12. Andrea Braides and Anneliese Defranceschi: *Homogenization of multiple integrals*
13. Thierry Cazenave and Alain Haraux: *An introduction to semi-linear evolution equations*
14. J. Y. Chemin: *Perfect incompressible fluids*
15. Guiseppe Buttazzo, Mariano Giaquinta, and Stefan Hildebrandt: *One-dimensional variational problems: an introduction*
16. Alexander I. Bobenko and Ruedi Seiler: *Discrete integrable geometry and physics*
17. Doina Cioranescu and Patrizia Donato: *An introduction to homogenization*
18. E. J. Janse van Rensberg: *The statistical mechanics of interacting walks, polygons, animals, and vesicles*
19. S. Kuksin: *Hamiltonian partial differential equations*
20. Alberto Bressan: *Hyperbolic systems of conservation laws: the one-dimensional Cauchy problem*
21. B. Perthame: *Kinetic formulation of conservation laws*
22. A. Braides: *Gamma-convergence for beginners*
23. Robert Leese and Stephen Hurley (eds): *Methods and algorithms for radio channel assignment*

Methods and Algorithms for Radio Channel Assignment

EDITED BY

Robert Leese

Smith Institute for Industrial Mathematics and System Engineering
and
St Catherine's College, University of Oxford

and

Stephen Hurley

Department of Computer Science, University of Cardiff

OXFORD
UNIVERSITY PRESS

OXFORD
UNIVERSITY PRESS

Great Clarendon Street, Oxford OX2 6DP

Oxford University Press is a department of the University of Oxford.
It furthers the University's objective of excellence in research, scholarship,
and education by publishing worldwide in

Oxford New York
Auckland Bangkok Buenos Aires Cape Town Chennai
Dar es Salaam Delhi Hong Kong Istanbul Karachi Kolkata
Kuala Lumpur Madrid Melbourne Mexico City Mumbai Nairobi
São Paulo Shanghai Taipei Tokyo Toronto

Oxford is a registered trade mark of Oxford University Press
in the UK and in certain other countries

Published in the United States
by Oxford University Press Inc., New York

First published 2002

A catalogue record for this title is available from the British Library

Library of Congress Cataloging in Publication Data

Methods and algorithms for radio channel assignment / Robert Leese and Stephen Hurley.
(Oxford lecture series in mathematics and its applications ; 23)
ISBN 0 19 850314 8 (hbk)
1. Radio frequency allocation–Mathematics. 2. Frequency spectra–Mathematical models. 3. Wireless communication systems–Mathematical models. 4. Algorithms. 5. Radio frequency allocation–Economic aspects. I. Leese, Robert. II. Hurley, Stephen. III. Series.

TK6552 .M48 2002 384.54′524–dc21 2002075716

ISBN 0 19 850314 8

10 9 8 7 6 5 4 3 2 1

Typeset by Newgen Imaging Systems (P) Ltd., Chennai, India
Printed in Great Britain
on acid-free paper by
T. J. International Ltd., Padstow

PREFACE

Over the last ten years there has been a rapid growth in the demand for wireless (radio) services, as part of the ongoing communications revolution. Radio spectrum is now the raw material for key segments of the economy. But it is a limited resource, and so government and industry are faced with the task of creating conditions for future economic growth through efficient management of the spectrum. The ever-increasing demand on the spectrum seems set to continue for several years to come, with new applications emerging all the time.

To meet the demand for spectrum resources, there are three basic avenues available: improvements in wireless and radio hardware, to use less spectrum and to release spectrum for other purposes; more effective and efficient use of the radio spectrum through improved assignment techniques; and the use of higher frequencies in the electromagnetic spectrum (above 70 GHz), to accommodate a wider range of services.

The collection of papers contained in this book contributes to the second avenue, that is the development of mathematical models and associated computational techniques to enable the effective and efficient use of the radio spectrum. Most of this material has its origins in invited presentations at a workshop on *Methods and Algorithms for Radio Channel Assignment*, held at the University of Oxford in April 1997, following a successful funding bid to the MathFit (Mathematics for Information Technology) initiative of the EPSRC (Engineering and Physical Sciences Research Council).

This book will be of interest to researchers and graduate students in mathematics and computing, and also to service providers and spectrum regulators. A good understanding of the issues involved in channel assignment and the associated computational techniques opens the way to effective and efficient use of the radio spectrum. Radio channel assignment has traditionally been a problem in engineering, but severe spectrum congestion has brought to the fore the computational and, more recently, the economic aspects. It is now truly a multidisciplinary undertaking.

We take this opportunity to thank all the authors for their contributions. We are also grateful to Henrik Blank, Laurence Green, David Hendon, and Michael Hodson of the UK Radiocommunications Agency for their support and interest, and also to Jim Finnie and Jim Norton for their keen involvement in the original Oxford workshop.

Cardiff Stephen Hurley
Oxford Robert Leese
March 2002

CONTENTS

CONTRIBUTORS

Stuart Allen
Department of Computer Science, Cardiff University
PO Box 916, Cardiff CF2 3XF

David Cohen
Department of Computer Science, Royal Holloway, University of London
Egham, Surrey TW20 0EX

Jan van den Heuvel
Department of Mathematics, London School of Economics
Houghton Street, London WC2A 2AE

Stephen Hurley
Department of Computer Science, Cardiff University
PO Box 916, Cardiff CF2 3XF

Peter Jeavons
Oxford University Computing Laboratory
Wolfson Building, Parks Road, Oxford OX1 3QD

Robert Leese
Smith Institute for Industrial Mathematics and System Engineering
PO Box 183, Guildford GU2 7GG

Colin McDiarmid
Department of Statistics, University of Oxford
1 South Parks Road, Oxford OX1 3TG

Derek Smith
Division of Mathematics, University of Glamorgan
Pontypridd CF37 1DL

Ryszard Struzak
Route du Boiron 45, CH-1260 Nyon
Switzerland

Roger Whitaker
Department of Computer Science, Cardiff University
PO Box 916, Cardiff CF2 3XF

1

BACKGROUND

Stephen Hurley and Robert Leese

1.1 Aspects of channel assignment

1.1.1 Engineering background

The field of research that is known by the generic term *frequency assignment* or *channel assignment* can be considered to have started in the 1960s. The basic requirement, because of spectrum scarcity, is the need for efficient *reuse* of the spectrum available for any given radio service. The basic prohibiting factor in radio spectrum reuse is interference, caused by the environment or by other radio transmissions. Interference can be reduced by deploying efficient radio subsystems (improvements in hardware) and by making use of better channel assignment techniques.

In radio transmission subsystems, techniques such as the deployment of time and space diversity systems, low noise filters, and the use of efficient modulation schemes can be used to suppress interference and safeguard the required signal. However, co-channel interference caused by frequency reuse is the most important constraint on system performance and capacity in radio networks. Therefore, the main motivation of channel assignment algorithms is to ensure that the carrier-to-interference ratio throughout the service region is acceptable, by making use of the propagation path loss information on the terrain under consideration.

The radio spectrum for a given service is generally divided into a set of disjoint radio channels. In order to divide a given portion of the radio spectrum into such channels, several techniques can be used, the most common being frequency division, time division, and code division. In frequency division, the spectrum is divided into disjoint frequency bands, whereas in time division the channel separation is achieved by dividing usage of the channel into disjoint time slots. In code division, the channel separation is achieved by using different modulation codes. More elaborate schemes can also be constructed; for example a combination of frequency division and time division would divide each frequency band of a frequency division scheme into time slots. This is the hybrid scheme generally used in current GSM telephony networks.

Channel assignment problems can be divided into several categories, for example fixed (or static) channel assignment (FCA), dynamic channel assignment (DCA) and hybrid channel assignment (HCA). In FCA problems, the service area is partitioned into a number of cells, and a number of channels are assigned to each cell such that some measure of interference is minimized. Such algorithms are relatively simple but cannot adapt to changing capacity requirements. In DCA, all channels are available and are assigned to new connections as needed such that a minimum quality-of-service threshold is satisfied. DCA schemes offer flexibility and adaptability but are more complex and are less efficient under high load conditions. Hybrid schemes combine elements of FCA and DCA. Other possible classifications distinguish whether assignment techniques are implemented in a centralized or distributed way. In the centralized versions, channels are assigned by a central controller, whereas in distributed versions channels are selected either by the local base station of the cell from which the connection originated or by the mobile receiver.

The technical chapters in this book assume frequency division, and centralized, fixed, channel assignment. These topics reflect the foundations of the channel assignment problem. Dynamic channel assignment, frequency hopping, and distributed techniques are more specialist topics, for which the interested reader is directed to the extensive research literature.

1.1.2 Overview of this volume

Although there is an ordering to the chapters, for the most part they do not depend on each other sequentially. Each chapter was written to be an independent contribution, with a small amount of cross-referencing between them. The subsequent chapters are arranged as follows.

Chapter 2 gives an introduction to the history of spectrum management from the viewpoint of the radio spectrum as a public resource. A historical discussion of the significance of the radio spectrum in society is followed by some recent examples of the impact of certain satellite systems, such as the Global Positioning System (GPS). Chapter 2 also discusses the role and mechanisms of international regulation, for example, how spectrum allocation takes place at national and international level.

Chapter 3 describes computational techniques for solving various forms of channel assignment problems, using so-called meta-heuristics. Meta-heuristics are a class of approximate methods that are designed to solve difficult combinatorial optimization problems. The chapter discusses the general characteristics of three of the most widely used meta-heuristic techniques: genetic algorithms, simulated annealing, and tabu search. Their performance on channel assignment problems is presented and related to best possible solutions for widely available benchmarks.

Chapter 4 presents lower bounds for channel assignment problems. By their very nature, the problems that arise in channel assignment are computationally difficult to solve; therefore, theoretical lower bounds are required to indicate how good a particular solution is, and whether it is capable of further improvement. Also, lower bounds can be used to assess the effectiveness of particular algorithms. The lower bounds presented in the chapter involve minimum span channel assignment, that is the question 'what is the minimum number of channels required to satisfy all interference constraints that define a particular problem?' Lower bounds for fixed spectrum channel assignment are discussed briefly, but this is an area in need of much more research effort.

Chapter 5 discusses the consequences for the minimum span channel assignment problem of taking a regular cellular (or lattice) structure for the locations of the radio transmitters. Regular layouts of this type are often used in the initial stages of network planning exercises. Results are obtained first in the presence of just *co-channel interference*, that is interference from sources using the same radio channel, and then in the presence of interference also from nearby channels. The possibilities are considered of relaxing the assumption of a lattice structure and of having a different demand for spectrum at each transmitter.

Chapter 6 considers an alternative model for channel assignment, based on higher-order constraints. Chapters 3, 4, and 5 all consider a channel assignment problem in terms of channel separation constraints between pairs of transmitters: the so-called *binary constraint* or *constraint matrix* model. However, depending on how a channel assignment is assessed, it may be important to consider interference from several transmitters simultaneously, not just from a single dominant transmitter as is the case in the binary constraint model. This leads to the higher-order constraint model for channel assignment. Chapter 6 also presents an introduction to the general area of constraint satisfaction.

Chapter 7 introduces the topic of cell planning (also known as network design) in radio networks and its importance in relation to channel assignment. In general, channel assignment algorithms are used to assign channels to transmitters, to minimize interference. However, the actual location and configuration of the base stations has a considerable effect on the ability to generate spectrally efficient assignments. The cell planning problem attempts to design spectrally efficient networks that satisfy service and capacity requirements.

Chapter 8 contains a framework that brings together economic and engineering aspects of spectrum management. By using an economic model for the competition between radio service providers, the benefits of different allocations can be calculated in terms of operator surplus and consumer surplus. In this way, spectrum allocations can be matched to the demand for services and the effects of spectrum fees can also be studied. This type of model represents a first attempt to quantify some of the key economic issues described in the following section.

1.2 Trends in radio spectrum management

1.2.1 Increasing usage of the spectrum

Increased demand for spectrum results largely from the trend to mobile communications. Radio is the only communications medium that can provide true mobility. In the telecommunications industry, customers who were previously content to communicate from fixed locations now want to be able to communicate on the move. There is also increasing demand for mobile data communications, as users discover the logistical advantages. In the not-too-distant future, it will become standard for laptop computers to be equipped with radio links, enabling users to transfer data and access the Internet while on the move. In addition, demand for more established mobile technologies, such as private business radio, continues to rise steadily.

At the same time, demand for fixed radio services also continues to grow strongly. Point-to-point links are an important component of telecommunications networks. Cables provide an alternative in some circumstances but there will continue to be a need for radio. There is a similar demand for point-to-multipoint links for 'Radio Fixed Access' (RFA, also known as 'radio in the local loop') to make the final connection between telecommunications networks and residential and business premises.

The final main component of increased demand is broadcasting, both terrestrial and satellite. Liberalization has led to greater competition and choice. The move to digital transmission will in due course release valuable analogue spectrum for new broadcasting use or other applications.

1.2.2 Tools for spectrum management

Spectrum shortages and congestion cause delay in the introduction of new services and deny users access to radio. If the full benefits are to be gained from the use of radio, it is essential that the various national and international agencies with responsibility for managing the spectrum have the necessary tools for that task. Historically, spectrum was managed by purely regulatory means, often on a first-come-first-served basis, subject to various technical criteria. The licence fees paid by service providers and radio users were set to do no more than recover administrative costs. This regime worked well while spectrum was plentiful, but is no longer adequate to cope with high demand.

Exclusive reliance on regulation, coupled with a disparity between licence fees and the intrinsic economic value of the spectrum, can have the following undesirable consequences.

1. Spectrum is not assigned to the highest-value use and investment decisions are distorted, leading to misallocation of resources.
2. Spectrum management decisions are imposed on the basis of incomplete information and uncertainty about future trends, instead of being market-driven.
3. Administrative procedures for changing spectrum allocations and assignments are slow, retarding desirable technical progress and market developments.
4. Users have little incentive to give up unused or underused spectrum or to invest in more spectrum-efficient technology or services.

5. Regulation imposes hidden but sizeable compliance costs and denies users the right to make informed choices about how best to meet their communications needs.

Many spectrum management agencies have now moved to more flexible spectrum management tools, based on various forms of 'spectrum pricing', in which market mechanisms allow licence fees to reflect more closely the economic value of the spectrum. To be most effective, these new tools should be be designed to meet key spectrum management objectives, such as balancing supply and demand for spectrum, promoting technical innovation and maintaining fair competition. They are not intended primarily as mechanisms for raising revenue.

The new market mechanisms do not replace spectrum regulation by any means. There remains the need for regulation, to implement international obligations; to define and implement overall spectrum strategy; to ensure fair competition; to maintain diversity in radio use, including by noncommercial users; and to deal with interference and illegal use.

The following paragraphs make a few very brief remarks by way of further background. There is plenty of further material that is readily available. For example, a very recent and very comprehensive review of spectrum management in the UK is contained in Cave (2002), which is available online at www.spectrumreview.gov.uk.

1.2.3 Spectrum pricing models

There are two broad models for spectrum pricing: *spectrum auctions*, where the market directly values the spectrum, and *administrative pricing*, where spectrum managers set a fee calculated to balance supply and demand. There is also a trend in several countries towards *tradable spectrum rights*, which effectively create a secondary market analogous to those for other commodities and raw materials.

1.2.3.1 Auctions Auctions have important advantages of economic efficiency, transparency, and speed. Compared to administrative pricing, they do not require the regulator to try to mimic the market or to guess market values, but allow users to value the spectrum directly. However, they are not suitable for all circumstances, especially where there are large numbers of relatively low-value licences to assign. In these circumstances, an auction would be unwieldy and transaction costs could exceed the value of the lots.

The design of the auction is crucial to a successful outcome. The simultaneous multi-round auction, pioneered by the Federal Communications Commission (FCC) in the US involves conducting the auction over an unlimited number of rounds with each bidder seeing the highest bid at each stage. It is particularly attractive where there are substitutable and inter-related lots that bidders may wish to combine in various ways, as it allows them to change strategy as the process unfolds and reduces the 'winner's curse' syndrome. The Internet has played a prominent role in the implementation of auctions and experiences from around the globe are readily available, including Australia (www.aca.gov.au), Canada (strategis.ic.gc.ca/SSG/sf01814e.html), New Zealand (www.med.govt.nz/rsm/), the UK (www.spectrumauctions.gov.uk), and the US (www.fcc.gov).

1.2.3.2 Administrative pricing Administrative pricing structures should be fair, transparent, and easily comprehensible by users, so that they send clear price signals. Various valuation methods have been proposed and there is no one-size-fits-all solution. By way of illustration, the approach adopted in the UK is that administrative pricing should be based on the least cost practicable alternative for improving spectrum efficiency.

1.2.3.3 Tradable spectrum rights In some countries, there has already been a shift from traditional apparatus licences, which define the technical characteristics of the radio equipment and the type of service to be offered, to flexible spectrum access rights. Such moves tend to be associated with secondary trading of spectrum, which can be a powerful agent for spectrum efficiency. Users have positive incentives to dispose of surplus spectrum through the market with a minimum of bureaucratic intervention, to others who can use it more productively.

Trading is generally seen as a very powerful tool for spectrum management, and also one that should be introduced with care. There could be problems with windfall gains and disorderly markets in moving directly from cost-based licencing to full secondary trading licences. Countries that have introduced secondary trading have generally done so only after auctioning the spectrum. Additionally, trading must be regulated in order to respect international agreements.

2

INTRODUCTION TO SPECTRUM MANAGEMENT

Ryszard Struzak

2.1 Introduction

2.1.1 Renaissance of radio

A hundred years after its invention, radio has entered a new era. The development of communication and computer technologies and their convergence generate applications that were difficult to imagine a few years ago. Radio has become indispensable for society to function. Areas in which radio waves have become irreplaceable are numerous: national defence, disaster warning, public safety, air-traffic control and weather forecasting are only a few examples. The most spectacular illustration is, however, the conquest of space; the moon landing in 1969 would never have been possible without radio (Dooling 1994). Remote sensing satellites are invaluable in discovering natural resources and monitoring the climate. With resolutions of the order of a metre available they allow the identification, quantification, mapping, and monitoring of land cover, agriculture, and water resources (Bell 1995). The world has seen an absence of major war conflicts thanks to 'spy satellites' that have continuously monitored military activities from space. Radio astronomy has opened new windows to the Universe and has contributed to a better understanding of the laws of nature (Zorpette 1995). Radio and television broadcasting have become the main source of everyday information for most people in the world. The Olympic Games, for instance, are watched by approximately two billion people, and there are more radio receivers than telephones throughout the world. Radio and television play a principal role in meeting the information needs of illiterate people.

Noncommunication applications of radio waves have also been developed. Millions of household microwave ovens are in daily use. Many industrial processes and scientific experiments have been improved, even made possible, by the ingenious use of radio waves (Struzak 1985).

Radio is now crucial for national and world economies, functioning like the nervous system in a living organism. The uses of radio waves create businesses. The enormous impact of radio on our lives continues to increase, although we still do not fully realize all the consequences of that development. In any event, it is widely accepted that the convergence of wireless telecommunications and information technologies will be a major engine for economic growth and improvement in standards of living during the decades ahead.

2.1.2 Application examples: GPS, Teledesic, and SPS

This section gives just three examples of applications of radio waves that could change our lives in the years ahead.

The first is the Global Positioning System (GPS). It is a space-based navigation, positioning, and time-transfer system completed in 1993 and offering unsurpassed accuracy, reliability, and availability. Now open for civilian applications at no cost, it was developed for military purposes at a cost of over US$10 billion. GPS is American; its Russian equivalent is named GLONASS. With a hand-held receiver costing about US$200, you may determine your position within an accuracy of 30 m. With special equipment, software, and access to the decryption key, this accuracy may be improved many times.

The operating principle is remarkably simple, and refers to the ancient art of navigation using stars in the sky. The main difference is that GPS uses 'artificial stars', that is a constellation of 24 satellites. Each satellite carries a precise onboard atomic clock. The position of each satellite is monitored by the GPS Master Control Station that also maintains a GPS time standard which, in turn, is synchronized with Coordinated Universal Time. The current satellite position and time, updated periodically, are uploaded to each satellite to be broadcasted continuously in a coded form. A GPS receiver extracts the data, and compares its own time with the time sent by a satellite. The difference between the two times is used to calculate the distance from the satellite to the receiver. The satellite clocks are exact to a billionth of a second (which corresponds to a 0.3 m uncertainty in distance); however, the receiver's clock is simple, to keep its weight and cost low, and introduces an unknown time offset. Thus, to calculate its longitude, latitude, altitude, and time offset, a GPS receiver must use data from at least four satellites. For this purpose, the satellites orbit in a formation that ensures that every point on the planet is always in radio contact with at least four satellites.

The precise time from the GPS satellites is used as a worldwide time reference to synchronize various clocks, generators, processes, and networks, including digital telecommunication and power networks, easily accessible from any point on Earth for the first time in history. But it is only one of the benefits available. The GPS system provides a unique address for each point on the Earth, instantly available in electronic form, setting a new standard for locations and distances. Its applications appear to be virtually unlimited. GPS enables drivers, mariners, and pilots to navigate safely and efficiently in all weather conditions, day and night, and to save fuel by travelling the most efficient route and at optimal speed. It provides data for mapping and surveying, laying roads, and building bridges, foundations, and utilities quickly and accurately. Once gathered, GPS data can automatically be transferred to a Geographic Information System (GIS). According to some predictions, GPS receivers may become as ubiquitous as watches, and

GPS coordinates may eventually replace a street address to define the location of a home or business.

The second major application is satellite communication services. Several constellations of new low-orbiting satellites are planned and have so far cost about US$8 billion. Among them, 'Teledesic',[1] licenced in 1997, is the most refined. It will provide affordable two-way communication services, such as broadband Internet access, videoconferencing, high-quality voice, and other digital data exchange, offering access speeds many times faster than today's standard modems. The Teledesic network is designed to support millions of simultaneous users, offering the same services everywhere on the planet.

The significance of these new communication applications cannot be overestimated. Information exchange and computer applications become increasingly important to economic development, education, health care, public services, and many other activities. However, most of the world does not have access even to the most basic telephone service. Even where this basic service is available, most of the networks over which it is provided are inappropriate for computer communications. They are antiquated and in need of modernization. The cost and the time required to upgrade these facilities through conventional lines would be prohibitive for much of the world. The new satellite systems are capable of providing the required services at a lower cost, indifferent to distance or location. Because satellites in polar orbits move in relation to the Earth, the cost of continuous coverage of any one point on Earth is the same as the cost of covering of all points on the Earth. It radically transforms the economics of telecommunications infrastructure and enables leapfrogging earlier stages of telecommunication development to gain immediate access to the most advanced information infrastructure. The value of such systems lies in the number of people gaining access to advanced communication services, and who otherwise will never have such access at all.

The third application example has nothing in common with communications, except for potential interference. The project, known as the Satellite Power System (SPS), was developed for the US government to satisfy its increasing energy demand. A constellation of 60 or so satellites would collect energy from solar radiation above the Earth's atmosphere, beam it to the Earth, and connect into the existing power distribution system. Each satellite would carry a solar array, 10 km long and 5 km wide each, to intercept approximately 70 GW solar energy. With a conversion efficiency of 7 per cent, this solar array would provide 5 GW of DC power for conversion to microwave power by thousands of klystron generators. This power would then be beamed to Earth via a 1 km diameter phased array microwave antenna. A special pilot beam would be transmitted

[1] In February 2002 Teledesic has signed an agreement with the Italian satellite manufacturer Alenia Spazio SpA to build the first two satellites. The system design has been changed to reduce cost. The new constellation design features 30 medium-Earth-orbit satellites. The first 12 satellites are expected to cost under US$1 billion, and will cover the densely populated areas of the world. The next 18 satellites will enable full global coverage capability.

from the Earth to the satellite for dynamic phasing of the transmit antenna. Radio waves would transport the energy from the geostationary satellite orbit to a terrestrial receiving antenna array. The receiving antenna would be a 10 km by 13 km array built of multiple half-wave dipole elements feeding diodes to directly convert radio frequency (RF) energy into direct current. Filters would be inserted between the dipole and diodes to suppress reradiation of harmonics generated in the rectification process. The system would cover a significant portion of national energy consumption and would cost a few billion US dollars. The project has not been implemented as the price of power from satellites would not be competitive and also because the environmental and EMC problems were inadequately addressed. If, however, the price of oil or carbon increases, the project may be revived.

2.1.3 Spectrum congestion

It is well known that applications of radio waves can interfere with each other and mutually nullify the benefits they offer. To avoid such interference, each application requires some amount of radio frequency spectrum for exclusive use, unless special arrangements can be made for sharing the spectrum with other applications. In the context of this chapter, we use interchangeably the terms 'radio waves', 'radio frequency spectrum', and 'spectrum'. The capacity that can be provided by any communication system to any single user, or to any group of users, is ultimately limited by the spectrum available to that system. The number of radio systems in operation worldwide is enormous and continues to increase. Liberalization and deregulation trends encourage the introduction of new services and new technologies, which generates demand for radio frequencies without precedence. The International Telecommunication Union (ITU) has recorded more frequency assignments in the last few years than during the whole previous history of radio. Most of the suitable frequencies have already been assigned and, within the existing arrangements, the demand exceeds what can further be assigned. In some frequency bands and geographical regions there is no room for new radio stations. Spectrum congestion is observed in the VHF/UHF frequency bands if the population density exceeds 200 people per square km and US$10,000 per capita per annum of GNP. Similarly, the geostationary satellite orbits have become congested and there may be no place for new satellites in some areas. This scarcity hampers further development.

Is spectrum congestion real and, if it is so, is there any way to solve it? Is the spectrum/orbit scarcity due to the laws of nature or, perhaps, due to our mismanagement? The issue is critical for the future of services and applications, and deserves serious consideration. It should be noted that the congestion of radio spectrum is not a new problem. In 1925 US Secretary of Commerce Herbert Hoover was the first to declare 'There is no more spectrum available' (Dougan 1992). In the meantime a multitude of applications of radio waves have been invented and successfully implemented. At the same time, the problem of the shortage of radio frequencies was repeatedly raised at international

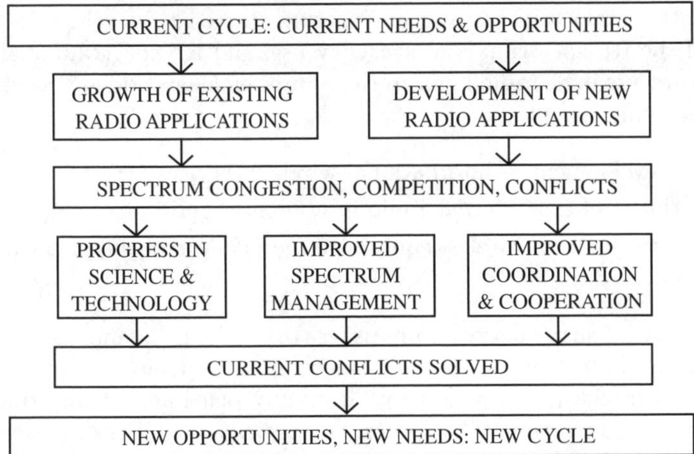

FIG. 2.1.

conferences and on other occasions, which indicates the periodicity of the process (see Fig. 2.1).

To solve spectrum congestion and radio interference problems, numerous conferences and symposia attract thousands of experts annually, and several international organizations have been involved. Among them, there are specialized UN Agencies such as the ITU, International Civil Aviation Organization (ICAO), International Maritime Organization (IMO), World Meteorological Organization (WMO), World Health Organization (WHO), World Trade Organization (WTO), and the World Bank. In Europe, the organizations involved include the European Union (EU), Conference of European Posts and Telecommunications Administrations (CEPT), European Radiocommunications Committee (ERC), and European Radiocommunications Office (ERO), among others.

2.2 Spectrum management

2.2.1 What is the spectrum?

What is the spectrum? The answer to this question is not as simple as one may expect. 'Spectrum' may have more than one appropriate meaning, as our approach to the spectrum changes.

Originally, the spectrum was only an abstract mathematical idea introduced by Jean-Baptiste Fourier (1768–1830) to solve differential equations. Initially the idea was strongly criticized and considered as a curiosity of doubtful value. Only when Peter Dirichlet (1805–59) and Georg Riemann (1826–66) resolved the doubts, was it generally accepted to be a powerful tool used in many branches of theoretical sciences. In the meantime, experimental science and instrumentation developed and the spectrum has also become a measurable physical object.

Consequently, radio sciences and engineering started to develop and now the concept of the RF spectrum is in everyday use, and RF spectrum analysers are basic instruments in all radio laboratories. Three elements added new dimensions to the spectrum concept:

(1) The development of international wireless services,
(2) the threat of cross-border radio interference, and
(3) the pressure from wireless equipment manufacturers, suppliers and service providers.

The ability to carry energy and messages over a long distance, with the speed of light and at no cost, made the radio spectrum a valuable resource from which everyone can profit. Free access to it, from any place and at any time, added much to its attractiveness. The spectrum has become a natural resource, with which another abstract quantity has been associated: the geostationary satellite orbit (GSO). An international treaty, signed by all governments, confirmed that '... radio frequencies and the geostationary satellite orbit are limited natural resources [...] that must be used [...] so that countries and groups of countries may have equitable access to both...' (ITU 1992). Radio waves and satellite orbits have become a common heritage of humanity (Fleming 1985). No one nation can operate them alone, ignoring others. They are subject to pollution by man-made radio noise and interference that decreases their utility.

2.2.2 Resource sharing

Common resources, such as radio frequencies and satellite orbits, have one disadvantage, best explained by Hardin, on a simplified model of a common pasture exploited by a group of herdsmen (Hardin 1968). He made three assumptions. First, he assumed that herdsmen are 'rational', that is each herdsman seeks to maximize his gain from the sale of animals, and there are no other rules regulating the use of the pasture. Second, the pasture is common and each herdsman pays nothing for grazing his herd there. Third, the pasture is limited. Under these assumptions, the scenario develops following the inherent logic of the commons, says Hardin. As each animal offers a unit gain, each herdsman tends to maintain as many cattle as possible. Initially, when the total number of animals is small, such an arrangement works well, and the herds and wealth grow quickly. However, when the population of animals approaches the carrying capacity of the pasture, adding more animals leads to degradation of the pasture and to its transformation to a desert. At that time, each herdsman asks himself what is the utility of adding one more animal to his herd. There are two components that he takes into account, one positive and one negative. The positive component is +1, since a herdsman receives all the proceeds from the sale of the additional animal. The negative component results from the additional overgrazing created by one more animal. Since, however, the effects of overgrazing are shared by all herdsmen, the negative utility for any particular decision-making herdsman is only a fraction of -1. Adding together these partial components, the herdsman concludes that the only reasonable course for him to pursue is to add another animal to his herd, then another, and another, and so on. And this is the conclusion reached by each and every rational herdsman sharing common pasture. Hardin concludes 'Therein is a tragedy. Each man is locked into a system that compels him to increase his herd without limit—in a world that is limited. Ruin is the destination toward which all men rush, each pursuing his own best interest in a society that believes in the freedom of the commons...' (Hardin 1968).

2.2.3 Regulation

The concept of free, unregulated access to a limited resource does not work if there are many users of such a resource. With radio waves, this conclusion was reached very early on. The first uses of radio were military, to communicate with warships at sea. Soon, however, military secrets were abandoned under the pressure of business exploiting the 'no man's land' of civilian radio. Two opposing forces appeared, one diverging and another converging. The diverging one was due to the competition among the equipment manufacturers and service providers. The converging force came from the market. The users of radio wanted to communicate freely with one another, independently of the service or equipment supplier. Moreover, in a liberal environment, without any coordination and control, mutual interference paralysed the operation of wireless systems. Finally, all interested parties came to the conclusion that coordination of activities was necessary. Such activities started on a national scale but, as radio waves do not recognize political borders, the global nature of the problem required international cooperation. Only two years after the first transatlantic transmission astonished the world, and just few years after Marconi received his patent on wireless telegraphy, the first international conference took place in Berlin in 1903 to regulate the use of the spectrum. The Berlin Conference marked the end of the first period of uncontrolled rivalry in radio communications. The spectrum resources consisted at that time of only two frequency bands (one near 500 kHz and another about 1 MHz) and were used to aid marine disaster relief (JTAC IRE-RTMA 1952).

The use of the spectrum has been regulated *by necessity*. There are two basic reasons for this: to prevent mutual interference, and to allow for intercommunication. In the early days of radio the second reason was more important. In 1902, Prince Henry of Prussia attempted to send President Theodore Roosevelt a courtesy message while crossing the ocean after his visit to the US. He was refused service because the shore station, operated by the Marconi Company, would not deal with a ship station of its German competitor. The Berlin Conference ruled that communication services with ships must be provided, regardless of the system used. However, international treaties are all part of a worldwide game nations agree to play following certain rules, and agreement is an inevitable ingredient here. Great Britain and Italy, where Marconi was exploiting his system did not agree and made reservations. Such reservations were removed, and the first operational standards on radio communications were agreed, only nine years later, at the London Conference in 1912. To a large extent, it resulted from pressure of public opinion shocked by the Titanic disaster. The Titanic was the most luxurious and largest ship at the time, and considered unsinkable. During its maiden voyage, it hit an iceberg and sank on the night of April 14–15, 1912 with about 1500 passengers losing their lives. Inquiries alleged that another liner was nearby and could have helped had its radio operator been on duty and thereby received the Titanic's distress signals.

Today, the uses made of the spectrum are internationally coordinated in the framework of the International Telecommunication Union. This Specialized Agency of the UN has, as its guiding principle, the avoidance of radio interference and assurance of rational use of the spectrum/orbit resources, taking into account the special needs of developing countries (ITU 1992). Its mission is accomplished by:

(1) Satisfying the specific needs of countries through World and Regional Radio Conferences, Radio Regulations, and relevant agreements and plans;
(2) Coordinating efforts to eliminate harmful interference between radio stations of different countries;
(3) Studies and recommendations on technical and operational matters by radio communication Study Groups and Assemblies.

Radio Regulations have the status of International Treaty; each government warrants that they are respected by everybody under its jurisdiction (ITU 1996). To keep pace with technological, political, and economic changes, the Regulations are periodically reviewed at competent World or Regional Radio Conferences. Figure 2.2 illustrates the process.

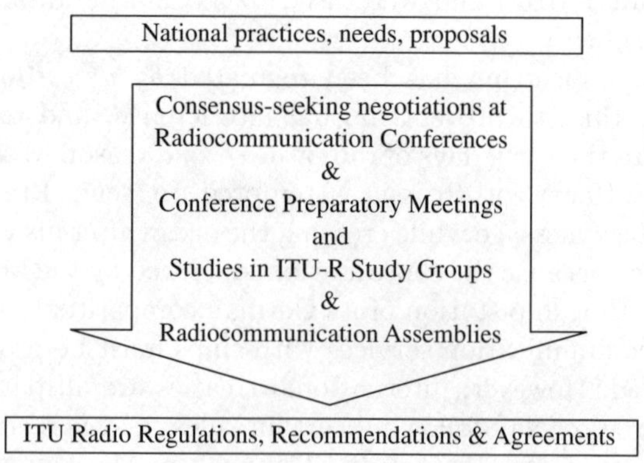

FIG. 2.2.

2.2.4 Evolution

The way we deal with spectrum or satellite orbit resources is evolving the same way as with other common resources. Previous experience confirms that the approach to such resources changes, as does our understanding of their value and social role. In uncontrolled growth, we have discovered with surprise that many resources, considered for a long time as being inexhaustible, have become scarce. Firstly, common land in food production was abandoned. Farmland has been enclosed, and now there is no free farmland. Later, open pastures and free hunting and fishing areas have been restricted. Also, using common land as a place for waste disposal has been abandoned and restrictions on the disposal of sewage are now widely accepted throughout the world. Finally, the concept of the environment and its protection has been developed and restrictions have been imposed on the pollution of land, water, and air. A short time ago, radio frequencies and geostationary satellite orbits entered the list of critical resources of humanity.

The issue of rational use, sharing, and protection of the limited common resources has become a serious problem on both the national and international scale. Several approaches are possible to solve it. 'We might sell them off as private property. We might keep them as public property, but allocate the right to enter them. The allocation might be on the basis of wealth, by the use of an auction system. It might be on the basis of merit, as defined by some agreed-upon standards. It might be by lottery. Or it might be on a first-come, first-served basis...' (Hardin 1968).

For the time being, on the international scene the spectrum/orbit resources are still treated as public, with the access rights acquired by negotiations among all country members of the ITU. However, on the national scene, most countries have introduced a system of fees. A few countries even created a spectrum market, but the market approach has not yet been universally accepted. There is a continuing debate over whether spectrum is to be treated as a free common heritage of humankind, a scarce natural resource, a renewable and reusable commodity, or a salable, auctionable, rentable piece of real estate. J. D. Bedin, a French jurist, defined it thus: 'the frequency spectrum is technology, industry, money, culture, and power' (Dougan 1992).

2.2.5 Preferences

Which of the possible approaches is the best? Each can be questioned, depending on the criteria applied. The final answer results from human preferences, goals and hierarchies of values. In practice, it is often impossible to separate technical aspects of resource sharing from their economic, social and political contexts, and from the interests affected by them. The problem of sharing scarce resources cannot be solved by technical means, without involving human values and ideas. Conflicting goals or values cannot be served by the same policies, since what enhances one will degrade the other. Our experience shows that the way in which the spectrum resources are used (and managed) follows the technological, economic, political and social changes in the world. These changes usually begin in one leading or dominating nation. Other nations, sooner or later, follow that example.

2.2.6 Objectives

National spectrum management began in the early 1920s with simple record keeping, that is logging out frequencies to applicants essentially on a first-come, first-served basis. The 1947 Atlantic City Radio Conference made foundations for the current international spectrum management system, by copying, to some degree, the US national system. Today, the concept of spectrum management embraces all activities related to planning, allocation, assignment, use, and control of the radio frequency spectrum and satellite orbits. To be effective, any spectrum management system should include monitoring and enforcement mechanisms and should be based on sound spectrum engineering.

Three objectives shape any spectrum management system: conveying policy goals, apportioning congestion, and avoiding conflicts, with due attention given to political, economic, ecological, and social aspects (Fig. 2.3). Society is composed of various groups, each with its particular interests, goals and views on how to use and manage the spectrum resources. As a consequence of spectrum congestion, conflicts arise between those who have access to the spectrum resource and those who do not. Conflicts arise also between the proponents of competing uses of the spectrum as well as between those who manage the spectrum and those who use it. These conflicts are of various types: commercial, political, physical interference, etc. (Huang 1993).

For those whose needs have already been satisfied, spectrum management should ensure the continuation of the existing status. Any modification would threaten their acquired benefits. On the other hand, newcomers have no access to the spectrum they need. For them, the principal aim of spectrum management

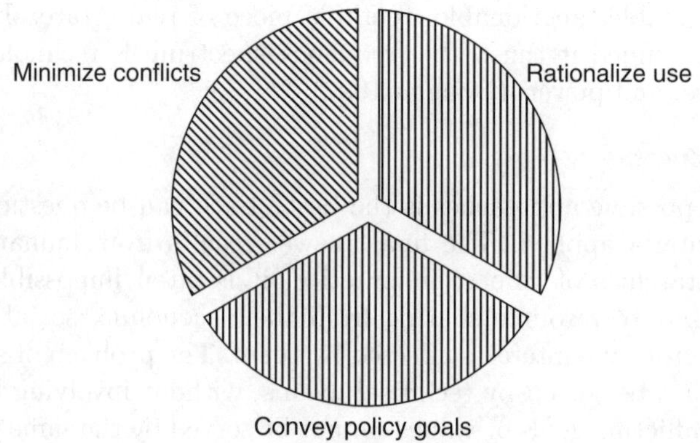

FIG. 2.3.

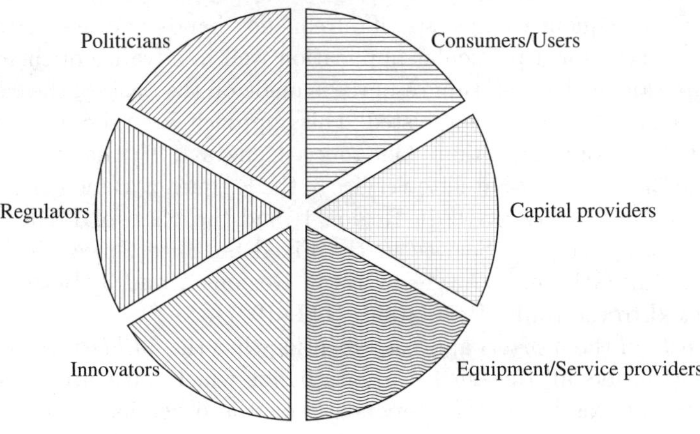

FIG. 2.4.

is to change the way the spectrum is assigned and to eliminate obstacles that prevent competition. What is best for one group is not necessarily good for the other. Since the very beginning, the spectrum management rules and regulations tend to reflect the relative balance of powers of the competing interest groups. Figure 2.4 shows the major players involved in spectrum management.

2.2.7 Dual approach

The uses made of the spectrum/orbit resources have been based on frequency allocation principles, as given in the Table of Frequency Allocations of the Radio Regulations. *Allocation* means the distribution of a frequency band to a wireless service, *allotment* means distribution to a country or area, and *assignment* means distribution to an individual radio station. Some allocations are worldwide, other are regional, that is uniform throughout a particular region. A country can make an assignment to an individual station or to a group of stations when needed. This is a so-called *ad hoc* frequency distribution method. An alternative is *a priori* frequency distribution. For services subject to *a priori* planning, an assignment in accordance with the plan receives protection from any other assignment. In the case of *ad hoc* managed services, the protection is given in accordance with the priority of registration dates—a system frequently described as first-come, first-served.

International frequency plans for specific applications, geographic regions, and frequency bands that are subject to *a priori* frequency planning are coordinated at competent radio conferences. A frequency plan is a table, or more generally, a function that assigns appropriate characteristics to each radio station at hand. The term 'frequency planning' remained from the early days of radio, when only the operating frequency of a radio station could vary. International plans are general and contain only a minimal number of details. In contrast, design and operational frequency plans include all the details necessary to operate the station.

In *a priori* frequency plans, specific frequency bands and associated service areas are reserved for a particular application well in advance of their real use. The distribution of the spectrum resource is made on the basis of the expected or declared needs of the parties interested. This approach was used, for instance, by the World Radiocommunication Conference, Geneva 1997, that established the plan for the Broadcasting-Satellite Service in the Frequency Bands 11.7–12.2 GHz in Region 3 and 11.7–12.5 GHz in Region 1, and the plan for Feeder Links for the Broadcasting-Satellite Service in the Fixed Satellite Service in Frequency Bands 14.5–14.8 GHz and 17.3–18.1 GHz in Regions 1 and 3. Both plans have been annexed to the Radio Regulations (ITU 1997).

Advocates of the *a priori* approach mention that the *ad hoc* method is unfair because it transfers all the burden to latecomers which must accommodate the requirements of existing users. Opponents, on the other hand, point out that *a priori* planning freezes the technological progress and leads to 'warehousing' the resources. Although all usable frequency bands have been allocated to services, only a small part of them is subject to *a priori* planning. In this regard, many developing countries are afraid that they will never have access to unplanned frequency bands. These bands will most probably be already occupied when the countries will be ready to use them.

Critics of *a priori* planning indicate that it is impossible to predict future requirements with a high degree of accuracy, and any plan based on unrealistic requirements has no practical value, blocks frequencies, and freezes development. Indeed, technological progress is very fast, and the plan may become outdated before it is implemented. In addition, there is no mechanism to limit the requirements, as the spectrum resource is available at no cost at international conferences. Although the ITU Convention calls for minimizing the use of spectrum resources, '... each country has an incentive to overstate its requirements, and there are few accepted or objective criteria for evaluating each country's stated need. In fact, the individual country itself may have only the dimmest perception of its needs over the time period for which the plan is to be constructed.... Under these circumstances, it is easy to make a case that allotment plans are not only difficult to construct, but when constructed will lead to a waste of resources as frequencies and orbit positions are warehoused to meet future, indeterminate needs...' (Robinson 1980). These remarks, however, do not concern frequency planning at the design stage of wireless systems, when all requirements are 'real'.

2.2.8 Trends

Current spectrum management policies and practices are inherited from the time when radio was mainly under state monopoly and access to the spectrum was free. However, the world has changed in the meantime. State monopoly is being abandoned and the importance of the private sector and non-governmental international corporations is increasing. A single market is being created and a competitive worldwide market economy is developing. New technologies are being created and introduced in the market at an accelerated rate. Digital signal processing has opened new horizons for the integration of services, for example, multiplexing, packed-switching, and spread-spectrum technologies offer new possibilities, not yet fully exploited. Satellite and stratospheric stations are being planned. All this does not fit into the framework of the present regulations. Redistribution, and better use of radio waves is felt necessary by many.

The present system has been criticized almost from the very beginning, but nothing better has been agreed on. The developing countries are afraid that there will be no spectrum to satisfy their future needs. They would also like to exploit their equipment while it is still usable. The developed countries are afraid that they cannot implement new technologies and develop new applications because of regulatory barriers. All users complain that the Radio Regulations are too complicated and excessively rigid (Bellchambers 1992). Every radio conference makes the participants equally unhappy with the results achieved. However, does the equal dissatisfaction of all parties involved not indicate that the best compromise possible has been reached? Otherwise, some will be more satisfied than others! Over the years, various improvements have been proposed, but few have been implemented. The fundamental rules remain unchanged. One of them has resulted in a separation of spectrum management from economic mechanisms. However, it is known from other fields that economic incentives could be used as an instrument to rationalize the use of scarce resources. When introduced internationally, spectrum occupation fees could limit excessive demand and warehousing of frequencies and orbital positions. Such proposals were discussed recently at the World Radiocommunication Conference, Geneva 1997, but never received substantial support. The majority of ITU members have preferred to continue with an administrative approach.

The concept of spectrum management through market forces was put forward by some countries, and has found as many supporters as opponents. The idea is to replace the regulatory mechanism by a competitive market economy mechanism. For the time being, this action has been limited to a few countries and to selected frequency bands only. Advocates of this idea indicate that market forces automatically match the demand to the available resource capacity and that market-based management is inexpensive. (The US Army alone spends nearly US$40 million each year in frequency compatibility investigations (Dougan 1992)). Moreover, relying upon administrative decision-making is inferior to relying on market forces because decisions are arbitrary and often mistaken in determining what is the best interest of users (Webbing 1977). Until now, however, no evidence has been published that selling the spectrum will solve the congestion problem in a way acceptable for all parties involved. Moreover, market forces could make wireless applications more expensive and influence the existing balance between the further developments of wired and wireless communication services.

An important event in recent years was a series of spectrum auctions conducted by the US Federal Communication Commission. These auctions mark a break with tradition. Earlier, licences to use the radio spectrum for wireless communication services were awarded on the basis of 'first-come first-served', by lottery, or by comparative hearings ('beauty contests'), almost for free. Now the federal agency is granting the licences to the highest bidders. The first auction in the USA held in 1994 concluded in assigning three 1-MHz bands around 900 MHz for a total of about US$650 million. In 1995, two pairs of 15-MHz bands around 1900 MHz for personal communication services were assigned for a total of US$7.74 billion (Bell 1996). In addition, the successful bidders have to pay expenses for relocating thousands of microwave transmission facilities that were already using that portion of the spectrum. These numbers, however, should not be generalized, as the price depends on supply and demand. Spectrum and real estate in the center of New York or Tokyo will cost much more than somewhere in a desert. However, one thing is clear, consumers will always pay the final costs.

The market approach combined with sovereignty, an indisputable principle in the ITU, may increase further the existing fragmentation in spectrum management. This fragmentation, manifested in differences in domestic regulations and legal provisions, may lead to a paradoxical situation such as that described by Zimmermann. If anywhere on the ocean a vessel with a crew of one is in distress, all related communications have absolute priority and are free of charge. (Such arrangements for distress signals from the sea have been agreed since the Titanic disaster, mentioned earlier.) However, if, after an earthquake, some 10,000 people are trapped under debris, any custom official can prevent the rescue team arriving from outside the affected country from bringing in their radios. And a different official can prevent them from using these radios, until they obtain a licence from the national spectrum management authority (whose building may just have collapsed in the earthquake). Also, if they are, nevertheless, able to use their satellite terminals, they may be presented with telephone bills for thousands of dollars. Such is the experience of those who provide international humanitarian assistance in the age of information superhighways, says Zimmermann (Zimmermann 1995).

2.3 Conclusions

How spectrum resources are used and managed has profound impact on society, on its prosperity, security, culture, and education. Following the tradition that originated at the Berlin and London conferences at the turn of the last century, the use and management of spectrum resources is based on few general principles, maintained with only minor changes until the present time:

1. The spectrum resources are public, and each country has equitable and free access to them. The uses of the spectrum resources are based on administrative regulations and allocations of frequency bands. These regulations and allocations are set through the mechanism of consultations, negotiations, and consensus.

2. Each sovereign country freely decides on its uses, as long as the electromagnetic compatibility principles and international agreements in force, including the Radio Regulations, are respected. Electromagnetic compatibility is the ability of a system to operate correctly in its environment without introducing intolerable disturbances to that environment.

There is growing opinion that the spectrum congestion is in a great part due to a combined effect of our inadequate approach, inappropriate regulations, simplistic models, tools and methods, and lack of reliable data. The existing spectrum management system needs to be improved and mechanisms to encourage spectrum conservation should be developed (Struzak 1993). Eventually, technology may remove the need for some functions now included in spectrum management. Future radio systems would be able to automatically coordinate among themselves the use of spectrum resources to avoid interference. However, in view of the enormous investment in 'old' equipment, it will not happen soon. Before this happens, wider application of advanced mathematical methods and computer techniques could increase the flexibility and objectivity of spectrum management and efficiency of spectrum use.

3

META-HEURISTICS AND CHANNEL ASSIGNMENT

Stephen Hurley and Derek Smith

3.1 Introduction

The channel (or frequency) assignment problem CAP (or FAP) arises in various forms and is a computationally hard problem to solve. In the 1960s and 1970s the effort was directed at developing simple greedy methods, heuristic methods (Zoellner and Beall 1977; Box 1978), and in very specialized cases, exact methods. During the 1980s general interest in so-called meta-heuristic methods dramatically increased. This was due to their apparent widespread success in solving a variety of difficult problems. This chapter will describe three such meta-heuristics: *evolutionary algorithms* (EA), *simulated annealing* (SA), and *tabu search* (TS). Their development for channel assignment will be discussed. Considerable interest has also been shown in using *neural networks* for channel assignment, but their performance has, at the time of writing, been relatively poor. They will, therefore, not be considered in detail here.

Meta-heuristics are a class of approximate methods that are designed to solve difficult combinatorial optimization problems. They provide general frameworks that allow for the creation of new hybrids by combining different concepts derived from classical heuristics, artificial intelligence, statistical mechanics, biological systems, evolution, and genetics. In particular, meta-heuristics use one or more of these concepts to drive some subordinate heuristic, such as local neighbourhood search.

Whereas specialized heuristics or optimization procedures can usually only be applied to a narrow band of problems, meta-heuristics can be applied to a wide variety of problems. However this robustness comes at the sacrifice of domain specificity. For example, if the function to be minimized is a quadratic bowl then Newton–Gauss optimization will generate the optimal point in one iteration from any starting location; it would be unlikely that meta-heuristics could compete in this instance. On the other hand, Newton–Gauss will generally fail to find the optima of multimodal surfaces because it relies heavily on gradient and higher order statistics of the function to be optimized. Meta-heuristics would generally have a much more satisfactory performance on such problems.

3.2 General search principles

The channel assignment problem can be modelled as a quadruple

$$(Z, D, C, g),$$

where

1. $Z = \{t_1, t_2, \ldots, t_N\}$ is a finite set of transmitters or base stations, that is objects to which a channel needs to be assigned.
2. $D = \{D_{t_1}, D_{t_2}, \ldots, D_{t_N}\}$ is a finite set of channel domains for the transmitters in Z. Often each transmitter has the same domain of F_{fixed} frequencies, $D = (d_1, d_2, \ldots, d_{F_{\text{fixed}}})$.
3. $C = \{c_1, c_2, \ldots, c_m\}$ is a finite set of constraints on a subset of transmitters in Z, usually identifying the required frequency separation between pairs of transmitters.
4. g is the objective function, which associates with every possible assignment a numerical value, usually corresponding to some interference measure.

The channel assignment problem is to assign a frequency f_i, where $f_i \in D_{t_i}$, to each t_i $(i = 1, 2, \ldots, N)$ such that the interference measure $g(f_1, f_2, \ldots, f_N)$ is minimized and (ideally) such that the constraints are satisfied.

Two approaches can be taken to solve this problem: either design a specialized scheme, that is one that is specific to the problem instance, or adapt some general existing scheme. If the latter approach is used then it is convenient to consider the following three main ways of searching the combinatorial spaces involved (Costa *et al.* 1995).

3.2.1 Constructive methods

This approach generates feasible solutions (assignments) by starting from an empty solution $s(0)$ and gradually assigning a frequency f_i to the current partial solution $s(i-1) = (f_1, f_2, \ldots, f_{i-1})$. There is a progressive reduction in the size of the search space X_i at each iteration, that is $X_N \subset X_{N-1} \cdots \subset X_1 \subset X$; see Fig. 3.1. The greedy method is an example of a constructive method. It simply takes the best choice at every iteration. Sequential assignment algorithms are greedy constructive methods which have been widely used in channel assignment (Hurley *et al.* 1997; Hale 1980). They consist of three main steps or *modules*. First, the transmitters are listed in some specified order and the first transmitter is assigned to the first available frequency. Second, the next transmitter to be assigned is selected (which may differ in some way from that implied by the initial ordering). Finally, the selected transmitter is assigned some frequency from the (ordered) frequency domain. Several options are available for the ordering used in each module, and therefore by using different options in each module a large number of different sequential methods can be obtained. For example, the FASoft system (Hurley *et al.* 1997) has a total of 64 different sequential assignment methods available.

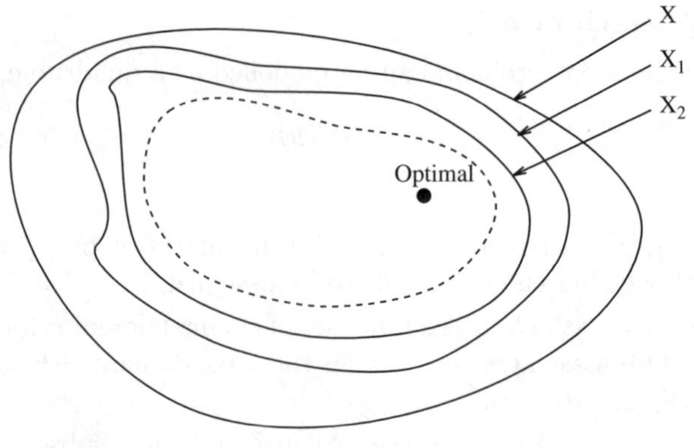

FIG. 3.1.

---

generate starting solution $s \in X$
repeat
  select $s' \in N(s)$ such that $g(s')$ is minimal
  if $g(s) > g(s')$ **then** $s \leftarrow s'$
until $g(s) < g(s')$

FIG. 3.2.

The advantage of constructive methods is that they usually produce solutions quickly, but often at the expense of poor solution quality.

3.2.2 Improvement methods

Improvement methods start with some initial, not necessarily feasible, solution s, which could be randomly generated or obtained from a constructive method, and involve selecting a neighbouring solution from the neighbourhood $N(s)$ of s. The neighbourhood $N(s)$ is the set of candidate solutions that can be reached from s by a simple operation σ. For example, σ could be the removal of an object from, or the addition of an object to, a solution. Simple local neighbourhood search, outlined in Fig. 3.2, illustrates the overall structure of the improvement approach. Figure 3.3 illustrates the operation of improvement methods. The starting point is $s(1)$, which is altered by some simple operation to give $s(2)$, that is $s(2)$ is within the neighbourhood of $s(1)$. In turn, $s(2)$ is then altered to give $s(3)$, which is within the neighbourhood of $s(2)$. This process is repeated until some termination criterion is satisfied.

This local search procedure will almost certainly terminate in a local minimum. Therefore, additional features are added in an endeavour to make further progress, giving rise to other improvement methods, such as simulated annealing and tabu search. The main difference between them is how they select the neighbouring solution at each iteration.

3.2.3 Evolutionary methods

Both the constructive and improvement approaches operate by considering one point in the search space (partial or complete). However, it seems reasonable to assume that performance could be improved by considering several such points in the search space, that is having a *population* of solutions, which can then be altered individually through *mutation* or in pairs through *recombination* or *crossover*. Each iteration or *generation* of an evolutionary algorithm usually involves both mutation and recombination. This is shown in Fig. 3.4, where the initial population consists of three candidate solutions (individuals), $s(1,1)$, $s(1,2)$, and $s(1,3)$. A new set of three individuals, $s(2,1)$, $s(2,2)$, and $s(2,3)$ is generated by mutating single individuals and by recombining certain pairs of individuals

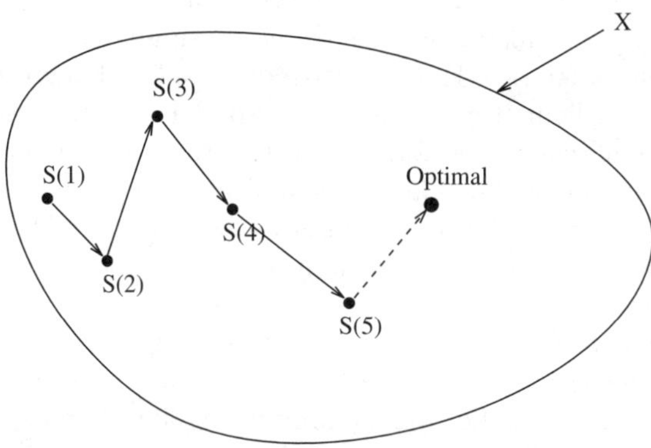

FIG. 3.3.

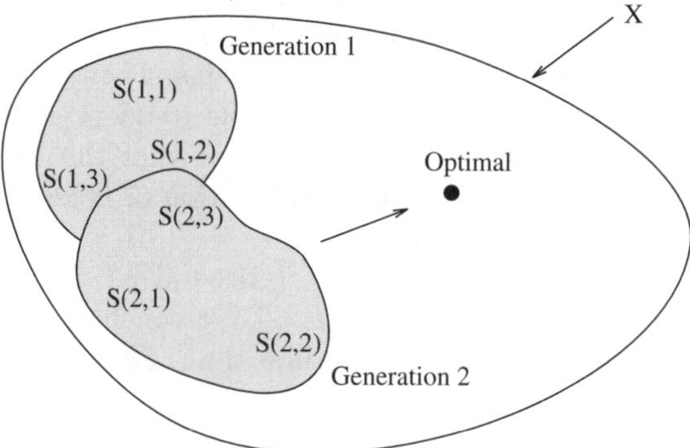

FIG. 3.4.

satisfying various conditions. This process proceeds until some termination criterion is reached. Such methods have the intuitive advantage that the search space is covered more widely at each iteration than in constructive or improvement methods. However, this can also be a drawback in that it often requires extra computational effort. Several different methods exist within this approach, for example, *evolutionary algorithms* (Back 1996), *scatter search* (Glover 1994) and *ant systems* (Costa and Hertz 1997).

3.3 Literature review

This section surveys the literature relating to the use of meta-heuristics in solving the channel assignment problem. It is not intended to be exhaustive but represents the majority of the research carried out so far.

Lanfear (1989) used a tabu search algorithm to solve the radio relay network frequency assignment problem. The problem was formulated as a generalized graph colouring problem and involved colouring several graphs which represented co-channel interference, first and second adjacent channel interference, and co-site interference. The algorithm was tested on simulated radio relay networks and air–ground–air scenarios. Results were compared with a greedy heuristic and a Hamiltonian path heuristic. It was observed that the tabu search algorithm performed well but was outperformed by the Hamiltonian path heuristic for the air–ground–air scenarios. This work indicates the importance of carefully considering the design of the meta-heuristic to achieve good performance and how an ineffective implementation usually leads to it being outperformed by a specialized heuristic.

Much work has been done with respect to solving cellular assignment problems using meta-heuristics. Duque-Anton *et al.* (1993) implemented a simulated annealing algorithm with the aim of minimizing interference while simultaneously assigning a certain prescribed number of channels to each cell. Mathar and Mattfeldt (1993) investigated the use of several algorithms based on the SA approach but using a different model. In a set of computational experiments the authors showed all variants gave good quality solutions when compared with optimal solutions obtained by tailored algorithms. Hao and Dörne (1996) used an evolutionary algorithm to solve cellular problems. They reported that the EA solved problems involving 300 cells, 30 available frequencies and 30,000 interference constraints. Empirical evidence suggested that the EA produced good results, but interestingly the effect of recombination or crossover was marginal, the dominant process being mutation.

Boyce *et al.* (1995) compared GENET (GEneral NETwork representation) and tabu search on the CELAR data set.[1] They concluded that GENET was more efficient in terms of computation time than TS, although this was due to extensive optimization of the algorithm. TS was found to perform better on small, tightly constrained problems; its weakness was that it employed no

[1] See ftp://ftp.cs.city.ac.uk/pub/constraints/csp-benchmarks.

techniques to keep track of neighbour evaluations and therefore a great deal of time was spent computing the cost function at each iteration. No description was given of how the short- and long-term memory were implemented. TS was able to find optimal solutions to some of the instances only when the complexity was reduced by using arc-consistency techniques. Kapsalis *et al.* (1995) also tackled the CELAR data sets, this time using an evolutionary algorithm. They concluded that the problem was very difficult for the basic EA with standard operators and attributed this to the high degree of interaction between links through both the constraints and specific domain requirements.

Several reports are available on the performance of meta-heuristics applied to the CELAR data sets, as a result of the CALMA (Combinatorial Algorithms for Military Applications) project.[2]

Hurley *et al.* (1996) reported on the use of meta-heuristics for solving a set of military fixed spectrum scenarios and found that the SA algorithm produced assignments with a smaller number of constraint violations than the assignments produced with either EA or TS. Castelino and Stephens (1999) implemented a tabu thresholding method that used surrogate constraints (SCTT) and reported that it outperformed tabu thresholding, simulated annealing, and evolutionary algorithms on fixed spectrum problems using simulated data sets.

Hurley *et al.* (1997) described a flexible system, FASoft, for solving channel assignment problems (fixed spectrum and minimum span). FASoft incorporates state-of-the-art heuristics (including SA, TS, and EA), sequential assignment algorithms, and a maximal clique algorithm to aid in the assignment process. Lower bounding techniques are included in the system to provide an assessment of how close the generated assignments are to optimal solutions.

As mentioned earlier, solution methods based on neural networks have in general underperformed when compared to other meta-heuristics or have performed well only on unrealistic small test problems. However, several algorithms have been presented. The interested reader can pursue this avenue in Kunz (1991), Funabiki and Takefuji (1992), Lochtie (1993), Lochtie and Mehler (1995) and Berger (1995). Neural networks have also been applied to dynamic channel assignment (Chan *et al.* 1991, 1994; Del-Re *et al.* 1996).

3.4 Overview of meta-heuristics

This section contains a brief overview of three types of meta-heuristics: evolutionary algorithms, simulated annealing, and tabu search. Comprehensive details of the methods can be found in Reeves (1995) and Aarts and Lenstra (1997). For a comprehensive account of application areas readers can consult Dell'Amico *et al.* (1997). Readers interested in the general principles of neural networks, which are not discussed here, can see Fausett (1994).

[2] See http://www.win.tue.nl/win/math/bs/comb_opt/hurkens/calma.html.

3.4.1 Evolutionary algorithms

Evolutionary algorithms are probabilistic optimization algorithms based on the process of natural evolution and population genetics. They rely on the collective learning process within a population of individuals, where each individual represents a point in the search space, that is a candidate solution. After an initial starting population is set up (either randomly or using domain-specific information), the population evolves towards better regions of the search space by means of the selection, mutation, and recombination (or crossover) processes. The optimization problem under consideration gives a measure of an individual's quality, that is its fitness, and the selection process is biased towards individuals of higher fitness surviving for the next generation of the evolution process. Recombination exchanges information between parent individuals and mutation introduces new information into the population.

They are three main types of evolutionary algorithms following these principles: *genetic algorithms* (Holland 1975; Goldberg 1989), *evolutionary programming* (Fogel *et al.* 1966; Fogel 1991, 1992) and *evolution strategies* (Rechenberg 1973; Schwefel 1977, 1981). An excellent account of all three types can be found in Back (1996). The main differences between them are in

(1) the representation used for individuals;
(2) the genetic operators used, that is recombination and/or mutation;
(3) the selection and reproduction scheme.

Evolutionary algorithms have the basic structure shown in Fig. 3.5 (with slight variations between types).

3.4.1.1 Representation or encoding mechanism It is assumed that a candidate solution to a problem may be represented by a finite set of parameters. These

---

$k \leftarrow 0$
initialize starting population $P(k)$
evaluate fitness of $P(k)$
while *not finished*
    select individuals from $P(k)$ to generate offspring $P'(k)$
    generate offspring using recombination and/or mutation
    evaluate fitness of $P'(k)$
    select $P(k+1)$ from $P(k) \cup P'(k)$
    $k \leftarrow k+1$
end while

FIG. 3.5.

parameters or *genes* are then joined to make a finite string or *chromosome*. The encoding mechanism depends on the nature of the problem variables. For example, when solving network flow problems the variables could assume continuous values, while the variables in a 0–1 decision problem may assume binary values. Evolution strategies are particularly suited to real-valued representations.

3.4.1.2 The fitness function A fitness function must be devised for each problem to be solved. Given a particular individual, the fitness function provides a single numerical value, which is a measure of the ability of the individual to solve the problem under consideration. For many problems, for example function optimization, it is obvious what this function should be, but for other problems things are not so clear, particularly in multi-criteria optimization problems.

3.4.1.3 Selection and reproduction To generate offspring, that is new candidate solutions, individuals are selected from the population and recombined. Parents are usually selected randomly, but biased towards fitter individuals. Good individuals will probably be selected several times, poor ones may not be selected at all.

Once individuals are selected for reproduction, recombination takes place. This usually involves *crossover* and/or *mutation*.

Crossover involves taking two selected individuals and combining them in some way to produce one or two new offspring. In the standard genetic algorithm, so-called one-point crossover is performed. A position in the chromosome is randomly determined as the crossover point, and an offspring is generated by concatenating the left substring of one parent with the right substring of the other parent. The second offspring is generated by concatenating the substrings not used to generate the first offspring. There are numerous extensions of this crossover operator, for example, two-point crossover and uniform crossover, and also more general recombination operators.

Whereas crossover involves using two individuals to produce new offspring, mutation is applied to single individuals, usually after crossover has taken place. Mutation in genetic algorithms was originally introduced as a background operator of little importance (Holland 1975). It works by changing gene values, for example by inverting bits in the case of binary alphabets, but with a very low probability, perhaps of the order of one altered gene in a thousand. Recent studies (Back 1993) have shown that much larger mutation rates, decreasing as the algorithm progresses, are often beneficial in helping the convergence, reliability, and speed of a genetic algorithm. Many other mutation operators, often problem specific, can also be defined.

3.4.2 Tabu search

Tabu search is an example of an improvement method as outlined in Section 3.2.2. It was first suggested by Glover (1977) and since then has become increasingly used. It has been successfully applied to obtain optimal or suboptimal solutions to such problems as scheduling, timetabling, travelling salesman, and layout optimization.

The basic idea, described by Glover *et al.* (1993), is to explore the search space of all feasible solutions by a sequence of *moves*. A move from one solution to another is the best available. However, to escape from locally optimal but not globally optimal solutions and to prevent cycling, at any iteration some moves are classified as forbidden or *tabu* (or *taboo*). Tabu moves are based on the *short-term* and *long-term* history of the sequence of moves. A simple implementation, for example, might classify a move as tabu if the reverse move has been made recently or frequently. Sometimes, when it is deemed favourable, a tabu move can be overridden. Such *aspiration criteria* might include the case where forgetting that a move is tabu leads to the best solution so far.

Formally, suppose g is the objective function on a search space X and it is required to find an $s \in X$ such that $g(s)$ has minimal value. For combinatorially hard (NP-complete) problems, this requirement is relaxed to finding an $s \in X$ such that $g(s)$ is close to the minimal value (a *suboptimal* value). This is because all known algorithms for determining the minimal solution require time which is exponential in the problem size. Suboptimal solutions may be found by halting when some threshold for an acceptable solution has been achieved or when a certain number of iterations has been completed.

A characterization of the search space X for which tabu search can be applied is that there is a set of k moves $M = \{m_1, \ldots, m_k\}$, the application of which to a feasible solution $s \in X$ leads to k (usually distinct) solutions $M(s) = \{m_1(s), \ldots, m_k(s)\}$. The subset $N(s) \subseteq M(s)$ of *feasible* solutions is known as the neighbourhood of s.

The method commences with a (possibly random) solution $s_0 \in X$ and constructs a sequence of solutions $s_0, s_1, \ldots, s_n \in X$. At each iteration, s_{j+1} is selected from the neighbourhood $N(s_j)$. The process of selection is first to determine the tabu set $T(s_j) \subseteq N(s_j)$ of neighbours of s_j and the aspirant set $A(s_j) \subseteq T(s_j)$ of tabu neighbours. Then s_{j+1} is the neighbour of s_j which is either an aspirant or not tabu and for which $g(s_{j+1})$ is minimal; that is, $g(s_{j+1}) \leq g(s')$ for all $s' \in (N(s_j) \setminus T(s_j)) \cup A(s_j)$.

Convergence to a local minimum can be avoided, since it is possible that $g(s_{j+1}) > g(s_j)$. The conditions for a neighbour to be tabu or an aspirant will be problem-specific. For example, a move m_r may be tabu if it leads to a solution which has already been considered in the last q iterations (*recency* or short-term condition) or which has been repeated many times before (*frequency* or long-term condition). A tabu move satisfies the aspiration criteria if, for example, the value of $g(s')$ for some $s' \in T(s_j)$ satisfies $g(s') < g(s_i)$ for all $0 \leq i \leq j$.

Example 3.1 Suppose for a positive integer B the search space is the discrete three-dimensional grid of B^3 vectors given by

$$X = \{(u, v, w) : 0 \leq u, v, w < B\},$$

and the objective function is $g(u, v, w)$. Consider the set of six moves which increment or decrement a single coordinate by one unit. Thus

$$M = \{m_1^+, m_1^-, m_2^+, m_2^-, m_3^+, m_3^-\} ,$$

with, for example, $m_1^+(u, v, w) = (u + 1, v, w)$. All moves are feasible moves except where s lies on the boundary of the grid. The inverse of move $m_j^\pm$ is $m_j^\mp$, for $1 \leq j \leq 3$.

A move is considered tabu if the inverse move has been made in the last q (e.g. 3) iterations or accounts for a fraction of more than p (e.g. 25%) of all previous moves. The former condition prevents short-term cycling and the latter long-term cycling. Both allow the search to descend from a local minimum. The aspiration criterion will be that a tabu move will be overridden if the solution becomes a new minimum over the history of the search.

Suppose $B = 7$, the initial solution is $s_0 = (3, 5, 2)$ and there have been 10 iterations with moves

$$m_2^+,\ m_3^+,\ m_1^-,\ m_2^+,\ m_1^-,\ m_3^-,\ m_1^-,\ m_2^-,\ m_2^-,\ m_3^+$$

Successive solutions $s_1, \ldots, s_{10}$, are therefore

$$(3,6,2), (3,6,3), (2,6,3), (2,7,3), (1,7,3)$$

$$(1,7,2), (0,7,2), (0,6,2), (0,5,2), (0,5,3) .$$

The following table analyses the possible candidate solutions s', one of which will be chosen as s_{11}:

Move	m_1^+	m_1^-	m_2^+	m_2^-	m_3^+	m_3^-
s'	(1,5,3)	(−1,5,3)	(0,6,3)	(0,4,3)	(0,5,4)	(0,5,2)
Comment		not feasible				tabu

Using the shorthand $g_i^\pm(s)$ for $g(m_i^\pm(s))$, the next move will correspond to the minimum of $g_1^+(s_{10})$, $g_2^\pm(s_{10})$ and $g_3^+(s_{10})$. Thus if $g_2^-(s_{10})$ is the minimum then $s_{11} = (0, 4, 3)$. However, if we also have $g_3^-(s_{10}) < g_2^-(s_{10})$ and $g_3^-(s_{10}) < g(s_j)$ for $j = 0, 1, \ldots, 10$, then the aspiration criterion forces the method to forget that move m_3^- is tabu and so $s_{11} = (0, 5, 2)$.

Example 3.2 In this example, the search space X is the set of all rooted binary trees with n terminal nodes (and therefore $n - 1$ internal nodes). The objective function g associates with each tree a real value and might, for example, involve the height of and the weighted distance between terminal nodes. The search commences with an initial tree whose nodes are labelled arbitrarily from 1 to $2n - 1$. A move m_{ij} consists of taking two nodes i and j and swapping the subtrees rooted at these nodes. Such a swap is valid only if i is not an ancestor

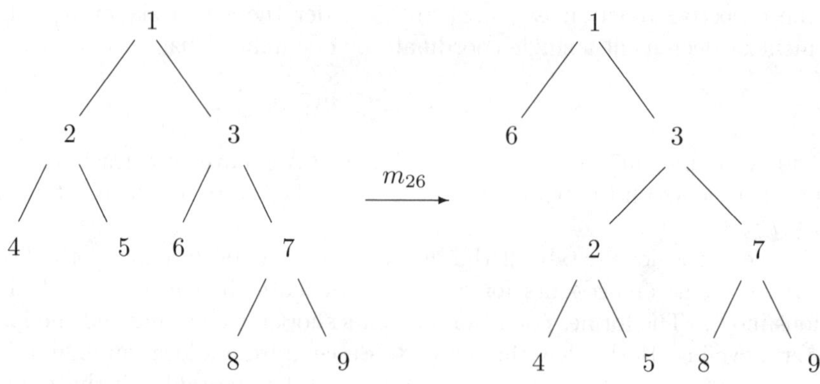

FIG. 3.6.

or descendant of j. Figure 3.6 illustrates the move m_{26} applied to a tree with 5 terminal nodes. Note that the set of terminal nodes is left invariant by such a move.

In this example, a move m_{ij} is tabu if i or j has been one of the nodes involved in a recent move. This representation allows the exploration of the whole search space from any initial tree. It has similarities with the successful implementation of tabu search for the travelling salesman problem in Glover (1991).

To implement a simple tabu search procedure for a particular application generally involves making the following steps:

Step 1: Define a *representation* for the problem, so fixing the structure of each point in the search space X, that is s in the above description.

Step 2: Define an *objective function* $g(s)$, which is to be maximized or minimized.

Step 3: Define a *neighbourhood structure*, $N(s)$.

Step 4: Define the *tabu rules* (the short- and long-term memory). These can be the simple recency or frequency rules previously mentioned or much more complicated rules such as those in Reeves (1995). The choice of rules defines the set $T(s)$.

Step 5: Define the *aspiration criteria*. As for the tabu rules, they can be simple as in Example 3.1 or more complex (again see Reeves (1995)). The choice of aspiration criteria defines the set $A(s)$.

Tabu search is a relatively simple technique. It can be applied in the basic form described above to a wide variety of problems. However, considerably improved results can be obtained by developing effective and efficient rules tailored to specific problems. Also, an apparent disadvantage of the method is that its efficiency can be dependent on fine-tuning a large collection of parameters. With tabu search, complexity is not only present in the application but also in the technique itself.

3.4.3 Simulated annealing

Simulated annealing (SA) is a stochastic computational technique derived from statistical mechanics for finding solutions of nearly minimum cost to large optimization problems. It is an example of an improvement method, discussed in Section 3.2.2, and can handle objective functions with many local minima. A procedure for solving optimization problems of this type should sample the search space in a way that has a high probability of finding optimal or near-optimal solutions in a reasonable time. Over the past fifteen years or so, simulated annealing has shown itself to be a technique which meets these requirements for a wide range of problems.

The method itself has a direct analogy with thermodynamics, specifically with the way that liquids freeze and crystallize, or metals cool and anneal. At high temperatures, the molecules of a liquid move freely with respect to one another. If the liquid is cooled slowly, thermal mobility is restricted. The atoms are often able to align themselves and form a pure crystal that is completely regular. This crystal is the state of minimum energy for the system, which would correspond to the optimal solution in a mathematical optimization problem. However, if a liquid metal is cooled quickly (quenched), it does not reach a minimum energy state but instead a somewhat higher state corresponding, in the mathematical sense, to a suboptimal solution.

In order to make use of this analogy, one must first provide the following elements:

(1) A description of possible system *configurations*, that is some way of representing a solution to the problem; usually each configuration s may be written $s = (x_1, x_2, \ldots, x_N)$ for some variables x_i;

(2) A *generator* of random changes to a configuration; these changes are typically solutions in the neighbourhood of the current configuration, for example, a change in one of the parameters x_i;

(3) An *objective* or cost function $E(s)$ (the analogue of energy), whose minimization is the goal of the procedure;

(4) A *control parameter* T (the analogue of temperature) and an *annealing schedule*, which indicates how T is lowered from high values to low values, for example, after how many random changes in configuration is T reduced and by how much?

Metropolis *et al.* (1953) first introduced these principles into numerical minimization. A succession of moves through neighbouring configurations is assumed to change its energy at each stage from E_{old} to energy E_{new} with probability

$$P(E_{\text{old}}, E_{\text{new}}) = \begin{cases} e^{-(E_{\text{new}}-E_{\text{old}})/kT} & \text{if } E_{\text{new}} > E_{\text{old}} \\ 1 & \text{otherwise}, \end{cases}$$

```
initialize T
generate random configuration s_old
calculate E_old = E(s_old)
while T > T_min
    for u = 1 to N_c
        generate new configuration s_new
        calculate E_new = E(s_new)
        take r uniformly at random from [0, 1]
        if E_new < E_old or r < P(E_old, E_new)
            s_old ← s_new
            E_old ← E_new
        end if
    end for
    reduce T (for example, T ← 0.9T)
end while
```

FIG. 3.7.

where k is a constant known as the Boltzmann constant. The connection with the Boltzmann distribution in statistical mechanics should be apparent. If $E_{\text{new}} < E_{\text{old}}$ then the new configuration has a lower energy than the old one and such a move is always accepted. If $E_{\text{new}} > E_{\text{old}}$ then the new configuration has a higher energy than the old one; such a move may still be accepted, but with a reduced probability, thus helping the system jump out of local minima. This general scheme, of always taking a downhill step while sometimes taking a uphill step is known as the Metropolis Algorithm.

The simulated annealing procedure in Kirkpatrick *et al.* (1983) uses the Metropolis Algorithm but *varies* the temperature parameter T from a high value (the 'melting point', where most new configurations are accepted) to a low value (the 'freezing point', where no new configurations are accepted). The full SA procedure for minimization is then as shown in Fig. 3.7 (for maximization set $E = -E$).

The algorithm can be viewed as a Markov chain on the space of all configurations; N_c is the number of steps taken at each temperature T, and is chosen with the aim of reaching a state of minimum energy before moving to a lower temperature.

3.5 Using meta-heuristics in channel assignment

It is usual to consider two types of problem within the model presented in Section 3.2: *minimum span* assignment and *fixed spectrum* assignment. In fixed spectrum assignment there is a fixed set of F_{fixed} channels. It is required to assign to each transmitter a channel from this fixed set such that (ideally) all the constraints are satisfied. In large problems it will be difficult, and often impossible to assign the channels in such a way, and therefore we look for an assignment in which as many as possible of the constraints are satisfied (or some other measure of interference is minimized).

In minimum span assignment, the set of available channels is not fixed, and could, for example, be represented by the set of consecutive integers $D = \{1, \ldots, K+1\}$. The requirement is to satisfy all the constraints in such a way that K is minimal over all feasible assignments. This minimum value of K is called the *minimum span*.

Of the three types of meta-heuristics presented in Section 3.4 we shall only consider here the detailed implementation of a simulated annealing and tabu search algorithm. Readers interested in using evolutionary algorithms can consult Crompton *et al.* (1993), Hao and Dörne (1996), and Crisan and Muhlenbein (1998).

Simulated annealing and tabu search are both examples of improvement methods and therefore iterate on a single candidate solution. Consequently, there are common components to the two algorithms, including the definition of a cost or objective function, the representation of an assignment, and the neighbourhood structures used. It will be seen later that both the fixed spectrum and minimum span implementations operate using a fixed number of channels, and therefore the following descriptions take F_{fixed} as the number of channels available, with a common domain $D = \{d_1, \ldots, d_{F_{\text{fixed}}}\}$ for all transmitters; F_{fixed} is constant in fixed spectrum assignment but varies in the minimum span assignment. Full details can be found in Hurley *et al.* (1997).

3.5.1 Objective function

The objective function g, minimized by both the SA and TS procedures, is formulated so that several factors can be taken into account:

(1) the number of violated constraints, e_{vio};
(2) the sum of the amounts by which each constraint is violated, e_{sum};
(3) the highest channel used, f_{large};
(4) the span, that is the difference between f_{large} and the smallest channel used, f_{small};
(5) the number of distinct frequencies used, e_{order};
(6) the largest of the constraint violations, l_{vio}.

Therefore, we choose as the objective function

$$g = \mu_1 e_{\text{vio}} + \mu_2 e_{\text{sum}} + \mu_3 (f_{\text{large}} - f_{\text{small}}) + \mu_4 e_{\text{order}} + \mu_5 f_{\text{large}} + \mu_6 l_{\text{vio}}\,, \quad (3.1)$$

where the μ_i are weights, which can include 0, reflecting the relative importance attached to the various terms. It is also possible to weight the different types of constraint (e.g., satisfying co-site constraints might be more important than satisfying adjacent channel constraints) or even individual constraints (e.g., to reflect the importance attached to certain areas of the overall network).

3.5.2 Representation and encoding

Let N be the number of transmitters to be assigned. A channel assignment $f = (f_1, \ldots, f_N)$ is represented using an array of indices $(x_1, \ldots, x_N)$, where $f_j = d_{x_j} \in D_j$ for $1 \leq j \leq N$. To illustrate this consider a problem which has six transmitters (numbered $1, 2, \ldots, 6$) and three channels available in a single domain $D = \{d_1, d_2, d_3\}$. The assignment $(2, 1, 1, 3, 3, 2)$ would indicate that transmitter 1 is assigned channel d_2, transmitter 2 is assigned channel d_1 and so on up to transmitter 6, which is assigned channel d_2.

3.5.3 Generation of neighbouring assignments

Several different move generators can be used to produce a new assignment from an old one. Note that SA and TS differ in that SA requires only a single neighbour at each iteration, whereas TS requires the complete neighbourhood (however defined) to be assessed.

3.5.3.1 Single move Here the neighbours of f are those assignments for which the array of indices differs in precisely one component, which is often randomly selected. Thus if f' is represented by $(x'_1, \ldots, x'_N)$ then $f' = (f'_1, \ldots, f'_N)$ is a neighbour of f if there exists j, $1 \leq j \leq N$, such that $x'_j \neq x_j$, and for all $i = 1, \ldots, N$ with $i \neq j$ we have $x'_i = x_i$. Tabu search would calculate the *full neighbourhood* of all $N(F_{\text{fixed}} - 1)$ possibilities, although in practice the neighbourhood is often truncated in some way.

Any assignment f has $N(F_{\text{fixed}} - 1)$ neighbours. Each neighbour corresponds to a pair (j, x'_j) with $1 \leq j \leq N$, $1 \leq x'_j \leq F_{\text{fixed}}$, $(x'_j \neq x_j)$. Double moves can also be defined in a similar way to the single move, the difference being that two indices are changed instead of one.

Since no restriction is imposed on which transmitter is chosen for reassignment, the single move generator could change the channel assigned to a transmitter which currently does not violate any intereference constraints. While this is not obviously a situation to be completely avoided, it can slow down the speed of convergence and also give unnecessarily large neighbourhoods in tabu search.

3.5.3.2 Restricted single move This generator and associated neighbourhood attempts to overcome the drawback of the single move, full neighbourhood. Let t_{v} be the number of transmitters in an assignment f which are assigned channels involved in one or more constraint violations. Denote such transmitters as *violating transmitters.* The *restricted single move* generator involves randomly selecting a violating transmitter i and setting $x_i = x'_i$ for some x'_i in the range $1, 2, \ldots, F_{\text{fixed}}$, either chosen randomly or to minimize some measure of interference.

The associated *restricted neighbourhood* consists of all neighbours of an assignment produced by the restricted move generator. For tabu search this involves searching $t_v(F_{\text{fixed}} - 1)$ neighbours and increases the efficiency of the algorithm compared to using the full neighbourhood, while still being effective.

To increase the efficiency of tabu search further, without sacrificing overall effectiveness, a *random restricted neighbourhood* can be used, defined as follows: the neighbourhood always consists of N neighbours (the restricted neighbourhood had t_v neighbours). Each neighbour is generated by randomly selecting a violating transmitter and randomly reassigning a different channel.

3.5.4 SA-specific aspects

3.5.4.1 Starting and finishing temperatures The starting temperature T_0 is determined by first setting $T_0 = 1$ and running the algorithm for N_c trials. If the acceptance ratio, χ, defined as the number of accepted trial assignments at the current temperature divided by N_c, is less than 0.9, double the current value of T_0. Continue this procedure until the observed acceptance ratio exceeds 0.9.

The SA algorithm terminates when the temperature falls below $T_{\min}$ (user specified) or the number of consecutive *frozen temperatures* exceeds a user specified value, usually 10. A frozen temperature occurs when no new assignments are accepted in the N_c steps making up the **for** loop in Fig. 3.7.

3.5.4.2 Annealing schedule There are many different annealing schedules. We have found the following two perform well.

The first (Costa 1993) uses a parameter α to govern the decrease in the temperature:

$$T_{k+1} = \alpha T_k, \quad \alpha \in (0, 1),$$

where k labels passages through the **while** loop in Fig. 3.7. The number of iterations N_c at each temperature is given by $N_0 = N$ (the number of transmitters) and

$$N_{k+1} = \max\left(2000N, \lceil N_k/\alpha \rceil\right).$$

As the temperature decreases, the number of trial assignments tested at each temperature increases, up to a maximum of $2000N$.

A more complex scheme, slower but often giving better results, reduces the temperature according to the scheme

$$T_{k+1} = T_k \left(1 + \frac{T_k N_c \ln(1+\delta)}{3\sqrt{\Phi}}\right)^{-1},$$

where δ is set to 0.1 and

$$\Phi = N_c \sum_{i=1}^{N_c} (g_i^k)^2 - \left(\sum_{i=1}^{N_c} g_i^k\right)^2 + 0.5,$$

with g_i^k being the value of the cost function for the ith trial assignment at the current temperature T_k.

3.5.5 TS-specific aspects

3.5.5.1 Short- and long-term memory A move to a neighbour (i, x_i') corresponding to changing the assignment of transmitter i to x_i' is said to be *tabu* if it does not satisfy the short- or long-term memory condition, defined as follows. Let x_i^j be the channel index assigned to transmitter i at iteration j. A move attempting to change x_i^k to some different index x_i^{k+1} is tabu (because of short-term memory) if, given a fixed integer $r > 0$, $x_i^{k+1} = x_i^t$ for some $t \in k-r, \ldots, k-1$. The move is also tabu (because of long-term memory) if, given a fixed $\beta \in (0, 1)$,

$$\frac{1}{k}\sum_{j=0}^{k-1} I\left[x_i^j = x_i^{k+1}\right] > \beta,$$

where $I[x_i^j = x_i^{k+1}]$ is an indicator function, taking value 1 if $x_i^j = x_i^{k+1}$ and zero otherwise. The short- and long-term memories are controlled by the parameters r and β, respectively. From computational experiments, their values are set to $r = \lceil 2N/5 \rceil$ and $\beta = 7/(4N)$.

3.5.5.2 Aspiration criteria Suppose the indices $\{x_i^j : i = 1, 2, \ldots, N\}$ represent the assignment f^j. A tabu move is selected if the neighbour f^{k+1} satisfies $g(f^{k+1}) \leq g(f')$ for all neighbours f' of f^k and $g(f^{k+1}) < g(f^j)$ for all $j = 0, 1, \ldots, k-1$.

3.5.5.3 Termination criterion The algorithm is terminated if zero interference has been obtained or if a predetermined maximum number of iterations is reached. The assignment with the lowest measure of interference is then taken as the final output.

3.5.6 Fixed spectrum procedure

There are several ways in which a fixed spectrum assignment algorithm can operate. The simplest way is to start with a randomly chosen assignment, in which transmitters are assigned channels randomly within the given spectrum, and then attempt to minimize some interference measure, for example the number of constraint violations. This approach has been used in Castelino and Stephens (1999). Alternatively, a constructive method and an improvement method can be combined. In this case, the procedure starts with a zero violation assignment, generated from a greedy sequential assignment method. Transmitters assigned channels outside the fixed spectrum available are then reassigned (possibly randomly) to channels within the allowed range. This, usually infeasible, assignment is then used as the starting point for an improvement method such as simulated annealing or tabu search. This approach is usually applied to the full problem, represented by the complete constraint graph (see Section 4.1). However it may

---

Step 1: Start with a zero violation assignment. If this has span p and $p < F_{\text{fixed}}$, the assignment is feasible, so stop.
Step 2: If $p \geq F_{\text{fixed}}$, reassign each transmitter previously assigned a channel above F_{fixed} with a randomly chosen channel in the allowed range.
Step 3: Use a SA or TS algorithm to minimize the interference measure g. After termination, output the best assignment found, which necessarily has span less than F_{fixed}.

FIG. 3.8.

Step 1: Generate a zero-violation assignment f^0, with span q.
Step 2: For each transmitter i with $f_i^0 = q + 1$, reassign f_i^0 randomly from the set $\{1, 2, \ldots, q\}$.
Step 3: Run an SA or TS algorithm, using channels $1, 2, \ldots, q$ and objective function (3.1) with $\mu_1 = \mu_3 = \mu_4 = \mu_5 = \mu_6 = 0$ and $\mu_2 = 1$, to eliminate any constraint violations introduced in step 2, that is attempt to find a zero-violation assignment with span $q - 1$.
Step 4: If a zero-violation assignment is produced in step 3, decrement q and go to step 2. Otherwise output q (the smallest span found with no violations) and the corresponding assignment.

FIG. 3.9.

also be applied initially to critical subgraphs, and then the resulting partial, possibly infeasible, assignment used as a starting point for the full calculation.

The general structure of fixed spectrum assignment algorithms is given in Fig. 3.8, where it is assumed that the channels $1, 2, \ldots, F_{\text{fixed}}$ are available at each transmitter.

3.5.7 *Minimum span procedure*

The general structure of the minimum span assignment algorithm is outlined in Fig. 3.9. It combines a constructive method (greedy sequential assignment) with an improvement method (either tabu search or simulated annealing). The starting point is any zero violation assignment.

All transmitters initially assigned the highest channel used are reassigned a randomly chosen lower channel. The previously highest channel is deleted from the set of channels available for use. This step generally introduces a small number of violations, but preserves most of the original assignment. The next step involves using SA or TS (the components of each algorithm are the same as in fixed spectrum assignment) to find a valid assignment with no violations using the reduced frequency domain. Once this is achieved, the assignment found is taken as the starting assignment and the process starts again. The loop terminates when the meta-heuristic is unable to reduce the number of violations to zero. The assignment taken as the starting assignment in the last iteration is then the best zero violation assignment found.

3.6 Improving assignment quality

3.6.1 Using subgraphs

The meta-heuristic algorithms described in the previous sections are usually applied to the full channel assignment problem under consideration. Although this is effective with a good choice of meta-heuristic, the quality of the assignment can sometimes be improved by using a two-stage procedure, concentrating first on a subproblem chosen to contain the essentially difficult part of the full problem. Once the subproblem has been successfully assigned, that part of the assignment is fixed and an attempt made to extend it to the full problem.

One possibility takes as subproblem a maximum level-p clique (for some p) in the constraint graph representation. A good choice of clique is one which gives the best lower bound of one of the types described in Chapter 4. The clique is assigned using one of the meta-heuristics; this partial assignment is then fixed and an attempt made to extend it to the full constraint graph G.

If not successful, this approach may possibly be improved by repeatedly adding carefully chosen vertices to the clique, so giving a sequence of subgraphs, along the lines discussed in Section 4.4.3. The illustrations presented below were obtained by adding vertices with the smallest reduced frequency domain. The two-stage assignment procedure is repeated with these subgraphs in place of the clique. Let t denote an acceptable difference between the span of the assignment of the full constraint graph G and the span of the subgraph assignment. If $U(X)$ denotes the subgraph of G induced by a set X of vertices, and W is the set of one or more vertices selected to add to the current subproblem C, then the procedure for minimum span problems is described in Fig. 3.10.

Ideally, the procedure terminates with $\sigma = 0$ and $s(G)$ equal to the best lower bound calculated from the current subproblem C using the methods of Chapter 4. In bad cases, $s(G)$ becomes larger than can be obtained by applying meta-heuristic methods to the whole problem directly.

If the clique or subgraph is not a significant determinant of the span, fixing its assignment may reduce the freedom of the meta-heuristic algorithm with no consequent benefit. Sometimes a useful compromise is to unfix the subgraph to restore its freedom, and hope that some good features of its assignment are retained when attempting to extend the assignment. If simulated annealing is used when doing this, the starting temperature should not be too high.

---

choose an initial clique C
assign C meta-heuristically, resulting in a span $s(C)$
extend the assignment to G, resulting in a span $s(G)$
$\sigma \leftarrow s(G) - s(C)$
do
  $\sigma_{\text{old}} \leftarrow \sigma$
  select a set W of vertices to add to C
  $C \leftarrow U(V(C) \cup W)$
  assign C meta-heuristically, resulting in a span $s(C)$
  extend the assignment to G, resulting in a span $s(G)$
  $\sigma \leftarrow s(G) - s(C)$
while $\sigma > t$ **and** $\sigma < \sigma_{\text{old}}$

FIG. 3.10.

Example 3.3 The Philadelphia problems are described in Example 4.10 (see also Fig. 4.2). They are based on a geometry originally used by Anderson (1973), and have subsequently been used as benchmarks by several authors (Gamst 1986; Sivarajan *et al.* 1989; Funabiki and Takefuji 1992; Leese 1996; Wang and Rushforth 1996).

An attempt to find a good assignment for the particular instance of Example 4.10 runs as follows. The maximum level-1 clique, C_1, has 186 vertices, corresponding to all the transmitters in cells 8, 9, and 16. An assignment of C_1 in span 413 can be found meta-heuristically. The lower bound for the span of C_1 obtained from the (PTMP + CAP) procedure described in Section 4.5.2 is 413. If an attempt is made to extend this assignment, the best span found is 451. A subgraph can be built from C_1 by successively adding the vertices with smallest reduced frequency domain, namely those vertices corresponding to cells 17, 2, and 15. The subproblem now has 275 vertices, and an assignment of span 426 can be found meta-heuristically, which can be extended to an assignment of the whole problem without increasing the span further, that is $\sigma = 0$ is reached in Fig. 3.10. Typically a direct application of meta-heuristic techniques without consideration of cliques gives an assignment of span 428 (see problem P1 in Table 3.2).

Example 3.4 As a second example, take the variation of the Philadelphia problem described in Section 4.5.3, originally used by Tcha *et al.* (1997). There is no need here to add vertices to the initial clique: the maximum level-0 clique, corresponding to cells $\{1, 2, 3, 7, 8, 9, 10, 15, 16, 17, 19, 20\}$, can be assigned meta-heuristically with span 524 and then extended to an assignment of the full problem with the same span, using a sequential algorithm. Typically a direct application of meta-heuristic techniques without consideration of cliques gives an assignment of span 530 (see problem P9 in Table 3.2).

3.6.2 *Philadelphia variations*

Additional variations on the Philadelphia problem can be formulated to further test the performance of the meta-heuristics and the improvement obtained from first assigning an appropriate subgraph. Full details can be found in Hurley *et al.* (1997) and Smith *et al.* (1998). Table 3.1 defines nine variations used in these papers (d_k denotes the smallest distance between transmitters which can use a separation of k channels, assuming the centres of adjacent cells are unit distance apart). The following demand vectors are used:

$$
\begin{aligned}
\boldsymbol{m_1} &= (8,25,8,8,8,15,18,52,77,28,13,15,31,15,36,57,28,8,10,13,8)\\
\boldsymbol{m_2} &= (5,5,5,8,12,25,30,25,30,40,40,45,20,30,25,15,15,30,20,20,25)\\
\boldsymbol{m_3} &= (20,20,20,20,20,20,20,20,20,20,20,20,20,20,20,20,20,20,20,20,20)\\
\boldsymbol{m_4} &= (16,50,16,16,16,30,36,104,154,56,26,30,62,30,72,114,56,16,20,26,16)\\
\boldsymbol{m_5} &= (32,100,32,32,32,60,72,208,308,112,52,60,124,60,144,228,112,32,40,52,32)
\end{aligned}
$$

Table 3.2 presents the results of applying the meta-heuristic techniques of Section 3.5. The lower bounds are found using the techniques described in Chapter 4. The remaining columns give: the span from the best sequential constructive method used in Hurley *et al.* (1997) (Best seq.); the best span produced by tabu search (TS); the best span produced by simulated annealing without pre-assigning critical subgraphs (SA); and the best span with SA when subgraphs are used in the assignment procedure (Using subgraphs). It can be seen that in eight out of the nine problems optimal solutions are found, with less than one per cent excess over the lower bound in problem P6.

TABLE 3.1.

Problem	d_0	d_1	d_2	d_3	d_4	d_5	Cell demands
P1	$\sqrt{12}$	$\sqrt{3}$	1	1	1	0	$\boldsymbol{m_1}$
P2	$\sqrt{7}$	$\sqrt{3}$	1	1	1	0	$\boldsymbol{m_1}$
P3	$\sqrt{12}$	$\sqrt{3}$	1	1	1	0	$\boldsymbol{m_2}$
P4	$\sqrt{7}$	$\sqrt{3}$	1	1	1	0	$\boldsymbol{m_2}$
P5	$\sqrt{12}$	$\sqrt{3}$	1	1	1	0	$\boldsymbol{m_3}$
P6	$\sqrt{7}$	$\sqrt{3}$	1	1	1	0	$\boldsymbol{m_3}$
P7	$\sqrt{12}$	$\sqrt{3}$	1	1	1	0	$\boldsymbol{m_4}$
P8	$\sqrt{12}$	$\sqrt{3}$	1	1	1	0	$\boldsymbol{m_5}$
P9	$\sqrt{12}$	2	1	1	1	0	$\boldsymbol{m_1}$

TABLE 3.2.

Problem	Lower bound	Best seq	TS	SA	Using subgraphs
P1	426	447	428	428	426
P2	426	475	429	438	426
P3	257	284	269	260	257
P4	252	268	257	259	252
P5	239	250	240	239	—
P6	178	230	188	200	179
P7	855	894	858	858	855
P8	1713	1800	1724	1724	1713
P9	524	592	530	538	524

$H \leftarrow G$
repeat
  calculate $P(v_i)$ for all $v_i \in V(H)$
  identify the set D of deficient vertices in H
  $H \leftarrow H \setminus D$
until $D = \phi$

FIG. 3.11.

3.6.3 J-kernels

Another method of determining a suitable subgraph is to use a *kernel*. Let ϕ_{ij} denote the constraint between vertices v_i and v_j in the constraint graph, so the assignment must satisfy $|f(v_i) - f(v_j)| > \phi_{ij}$. Choose a suitable target value J, normally some estimate of the span of G.

Definition 3.5 *Let*

$$P(v_i) = \sum_{v_k \in \Gamma(v_i)} (2\phi_{ik} + 1)$$

where $\Gamma(v_i)$ is the set of vertices adjacent to v_i in the constraint graph G. If $P(v_i) < J + 1$ then v_i is called a deficient vertex in G.

Deficient vertices are those for which a frequency is necessarily available if $G \setminus \{v_i\}$ is assigned using channels $\{0, 1, 2, \ldots, J\}$. (Here, in a slight abuse of notation, $G \setminus \{v_i\}$ is the induced subgraph of G obtained by removing the vertex v_i.) A *J-kernel* is a subgraph containing no deficient vertices, obtained as shown in Fig. 3.11.

For some choices of J no vertices are removed, and for some choices all vertices are removed. There is usually a range of values of J for which the J-kernel is a proper subgraph of G, encapsulating the difficulty of the entire problem. In this case the removed vertices are essentially easy to assign when extending an assignment of the J-kernel.

As with the clique-extension methods this approach is very successful for some problems, giving better assignments than applying the meta-heuristic to the full problem, and obtaining them more quickly. Generally, the larger the kernel the more certain it is that an assignment can be extended. However, for some problems the assignment of a large kernel appears to be no easier than assignment of the full problem. For the technique to be valuable, it must be possible to choose J so that both assignment of the kernel and the extension of the resulting partial assignment are relatively easy.

3.7 Summary

The channel assignment problem is a computationally hard problem of practical importance. Considerable effort has gone into devising methods for its solution that generate high quality assignments. The meta-heuristics presented in this chapter, particularly ones based on tabu search or simulated annealing and used in conjuction with constructive sequential methods, have proven themselves over a wide range of large realistic minimum span and fixed spectrum problems. However, the implementation of any meta-heuristic is critical to its effectiveness. The size of practical channel assignment problems is constantly increasing. It is therefore important to develop efficient and effective neighbourhood structures, and to choose carefully the parameters specific to each meta-heuristic.

In addition to combining improvement methods and constructive methods in the assignment algorithm, a further useful strategy, giving in many cases optimal results, is to first assign a subgraph of the constraint graph and then use it as a starting assignment (either fixed or unfixed) for the assignment of the whole problem. These subgraphs can be based on either cliques or J-kernels.

4

LOWER BOUNDS FOR CHANNEL ASSIGNMENT

Derek Smith, Stuart Allen, and Stephen Hurley

4.1 Introduction

The study of lower bounds for channel assignment (Raychaudhuri 1985; Gamst 1986; Lanfear 1989; Smith and Hurley 1997; Allen *et al.* 1999) is important for at least three reasons. First, lower bounds can be used to show how good a particular assignment is, and whether it is capable of improvement. Second, they can be used to assess the effectiveness of particular algorithms. This is a much more powerful technique than simply comparing the results of algorithms one with another. Indeed, it is unfortunate that many papers on channel assignment algorithms have simply given results for a particular algorithm on sets of test data which may not be accessible to the reader. This can make it impossible for the reader to assess the effectiveness of the algorithm. Third, it may be that identification of the particular structure in the problem that determines the best lower bound shows where the algorithm should concentrate its effort in order to find the best assignment.

This chapter will concentrate on lower bounds for minimum span channel assignments. The study of lower bounds for the number of constraint violations in fixed spectrum problems with no feasible solution is much less well developed. Only a brief discussion of this topic is included here.

Express the channel assignment problem in its usual graph theoretic formulation. The transmitters are represented by the vertices $V(G)$ of a constraint graph G (Hurley *et al.* 1997):

Definition 4.1 *A constraint graph G is a finite, simple, undirected graph in which each edge v_iv_j $(v_i, v_j \in V(G))$ has a non-negative integer label ϕ_{ij}.*

Definition 4.2 *A* channel assignment *(or frequency assignment) in a constraint graph G is a mapping $f : V(G) \to F$ (where F is a set of consecutive integers $0, \ldots, K$) such that the constraints*

$$|f(v_i) - f(v_j)| > \phi_{ij}$$

are satisfied for all $v_iv_j \in E(G)$. Sometimes this is referred to as a zero-violation *assignment. If one or more of the inequalities are violated then f is an assignment* *with constraint violations. The elements of the set F are referred to as channels (or frequencies).*

Definition 4.3 *If K is a minimum over all zero-violation assignments then the assignment f is a minimal assignment. This minimal value of K is the* minimum span *of G, denoted* $\mathrm{sp}(G)$.

Thus, the span of an assignment is the difference between the largest channel used and the smallest channel used and $\mathrm{sp}(G)$ is the minimum span over all possible assignments. The object of this chapter is to find a lower bound for $\mathrm{sp}(G)$, that is a value B such that $\mathrm{sp}(G) \geq B$. Then $B+1$ is a lower bound on the number of consecutive channels that could possibly be used to assign G without constraint violations. Sometimes an assignment of span B can be found; sometimes no assignment of span B is possible. A lower bound B is only useful if B is close to $\mathrm{sp}(G)$.

Proposition 4.4 *If G'' is a subgraph of a constraint graph G then* $\mathrm{sp}(G) \geq \mathrm{sp}(G'')$.

Proof Suppose $\mathrm{sp}(G) < \mathrm{sp}(G'')$. Given a minimum span assignment of G, the channels assigned to the vertices of G'' form an assignment of G'' of span smaller than $\mathrm{sp}(G'')$, which would give a contradiction. □

It follows that if B is a lower bound for $\mathrm{sp}(G'')$, then B is also a lower bound for $\mathrm{sp}(G)$.

4.2 Clique bounds

The most common lower bound is based on cliques. The idea is borrowed from the theory of graph colouring. A *clique* of a graph G is a maximal complete subgraph of G. Thus, every pair of vertices of the subgraph are adjacent, and the subgraph is not contained in any larger such subgraph (some authors omit this maximality requirement). For the purposes of channel assignment, cliques can be regarded as sets of transmitters for which there is a constraint between the frequencies assigned to any pair, so they tend to correspond to clusters of geographically close transmitters. As the edge labels of the constraint graph normally take several different values, it is possible to define several different levels of clique.

Definition 4.5 *A* level-p clique *of G is a complete subgraph in which every edge has label at least p, and which is not contained in any larger such complete subgraph.*

Thus a level-1 clique, for example, corresponds to a set of transmitters in which no pair can be assigned the same channel, or a first adjacent channel.

Theorem 4.6 *If C_p is a level-p clique of a constraint graph G then*

$$\text{sp}(G) \geq (p+1)(|V(C_p)|-1)$$

where $V(C_p)$ denotes the vertex set of the clique C_p.

Proof By Proposition 4.4, the minimum span of G cannot be less than the minimum span of the subgraph C_p of G. For any chosen minimal span assignment f of C_p renumber the vertices of C_p as $v_0, v_1, \ldots, v_{|V(C_p)|-1}$ in ascending order of the channel assigned to them. The span of the assignment is the difference between the largest and the smallest channel used, that is

$$\begin{aligned} \text{sp}(C_p) &= f(v_{|V(C_p)|-1}) - f(v_0) \\ &= \sum_{j=0}^{|V(C_p)|-2} (f(v_{j+1}) - f(v_j)) \\ &\geq \sum_{j=0}^{|V(C_p)|-2} (p+1) \\ &= (p+1)(|V(C_p)|-1). \end{aligned}$$

□

Example 4.7 The constraint graph shown in Fig. 4.1 has minimum span 11. A minimum span assignment is shown, with each tick representing a possible channel and the numbers above giving the vertex or vertices to which each one is assigned. Theorem 4.6 applied to the level-3 clique $\{2, 3, 5\}$ gives $\text{sp}(G) \geq 8$. Similarly, applying the theorem to the level-2 clique $\{2, 3, 4, 5\}$ gives $\text{sp}(G) \geq 9$. It will be seen later that the clique bound is capable of improvement for this example.

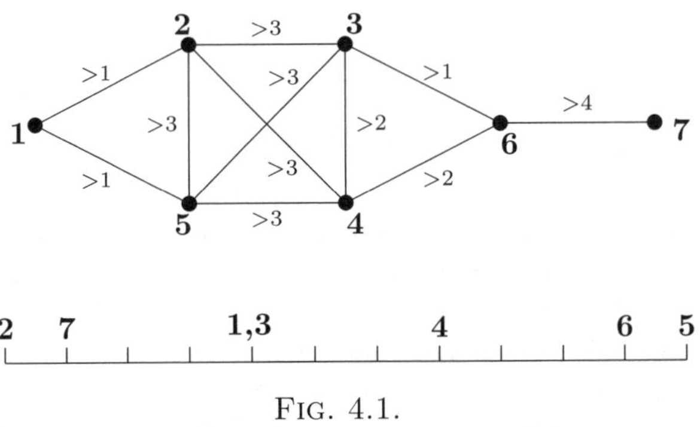

FIG. 4.1.

4.3 Travelling salesman bounds

A better bound can often be obtained using Hamiltonian paths. A Hamiltonian path in a graph G is a path through all of the vertices of the graph, visiting each vertex once and once only. Hamiltonian paths were first used to derive bounds for the channel assignment problem by Raychaudhuri (1985). An account is also given by Smith and Hurley (1997). In order to calculate the bound it is first necessary to construct from G a weighted complete graph G' on the vertices of G. The weight c_{ij} of each edge v_iv_j of G' is given by

$$c_{ij} = \begin{cases} 0 & \text{if } v_iv_j \text{ is not an edge of } G, \\ \phi_{ij}+1 & \text{if edge } v_iv_j \text{ has label } \phi_{ij} \text{ in } G \ (\phi_{ij}=0,1,\ldots). \end{cases}$$

The weight $W(H)$ of a Hamiltonian path H is the sum of the weights of all its constituent edges. Let $H(G')$ be the total weight of a minimum weight Hamiltonian path in G'.

Theorem 4.8 *If G is a constraint graph then*

$$\mathrm{sp}(G) \geq H(G').$$

Proof For any minimal span assignment f of G, number the vertices of G as $v_0, v_1, \ldots, v_{|V(G)|-1}$, in ascending order of the channel assigned to them (and arbitrary order for vertices assigned the same channel). Then $v_0, v_1, \ldots, v_{|V(G)|-1}$ is a Hamiltonian path H_f in G'. The span of the assignment is the difference between the largest and the smallest channel used, that is

$$\begin{aligned} \mathrm{sp}(G) &= f(v_{|V(G)|-1}) - f(v_0) \\ &= \sum_{j=0}^{|V(G)|-2} (f(v_{j+1}) - f(v_j)) \\ &\geq \sum_{j=0}^{|V(G)|-2} c_{v_{j+1}v_j} \\ &= W(H_f) \\ &\geq H(G'). \end{aligned}$$

□

The problem of determining the minimum weight of a Hamiltonian path in a graph is usually known as the open, symmetric travelling salesman problem (Volgenant and Jonker 1982). Therefore, the bound of Theorem 4.8 can naturally be referred to as the *travelling salesman bound.*

It is important to realize that Theorem 4.8 should be applied to a subgraph of the constraint graph and not to the constraint graph itself. In most problems the application of the theorem to the full constraint graph gives a lower bound which is too small to be useful. Instead the theorem is applied with G taken to be the chosen subgraph of the full constraint graph. In order to obtain a strong bound, a suitable choice of subgraph is a clique or a clique with some vertices added. The question of how this subgraph should be obtained will be considered later.

A method due to Volgenant and Jonker (1982) and software described by Volgenant (1990)[1] can be used to calculate the travelling salesman bound. The results are generally satisfactory. Even when the algorithm does not converge in reasonable time, a good lower bound for $H(G')$ is usually obtained, which of course is also a lower bound for $\mathrm{sp}(G)$. However, the software is only applicable for subgraphs of up to 250 vertices, which does sometimes restrict its application.

Another bound, which is easier to calculate, is the *spanning tree bound.* A spanning tree in a graph is a connected subgraph which contains every vertex and no cycles. Let $S(G')$ denote the total weight of a minimum weight spanning tree in G'. As a Hamiltonian path is a spanning tree, it follows that $H(G') \geq S(G')$. Thus, the following result is immediate:

Theorem 4.9 *If G is a constraint graph then*

$$\mathrm{sp}(G) \geq S(G').$$

Again the theorem should be applied to a suitable subgraph and not normally to the full constraint graph. The spanning tree bound is often not as strong as the travelling salesman bound, but is much easier to calculate. When applied to a clique it may be stronger than the clique bound. A simple greedy algorithm, such as Prim's Algorithm (Prim 1957) finds $S(G')$ in time $O(|V(G)|^2 \log |V(G)|)$. An implementation is contained in FASoft (Hurley *et al.* 1997).

Returning to Example 4.7 it can be seen that applying both the spanning tree bound and the travelling salesman bound to the clique $\{2, 3, 4, 5\}$ give lower bounds of 11, which are tight for $\mathrm{sp}(G)$.

Example 4.10 The Philadelphia problems were first described by Anderson (1973) and have been studied by many authors, see for example Gamst (1986) and Sivarajan *et al.* (1989). Optimal solutions to most of the standard variations have recently been found (Smith *et al.* 1998; Hurley *et al.* 1997). The problems are based on the area around Philadelphia, Pennsylvania. They have a geometry of 21 hexagonal cells, shown in Fig. 4.2.

A requirements vector $\boldsymbol{m}$ is used to specify the demand for frequencies in each cell. Thus m_i denotes the number of channels required by cell i and $\boldsymbol{m} = (8, 25, 8, 8, 8, 15, 18, 52, 77, 28, 13, 15, 31, 15, 36, 57, 28, 8, 10, 13, 8)$ in this example. Transmitters are considered to be located at cell centres and the distance between

[1] The software is available on the World Wide Web. See:
http://www.mathematik.uni-kl.de/~ wwwwi/WWWWI/ORSEP/contents.html
ftp://www.mathematik.uni-kl.de/pub/Math/ORSEP/VOLGENAN.ZIP.

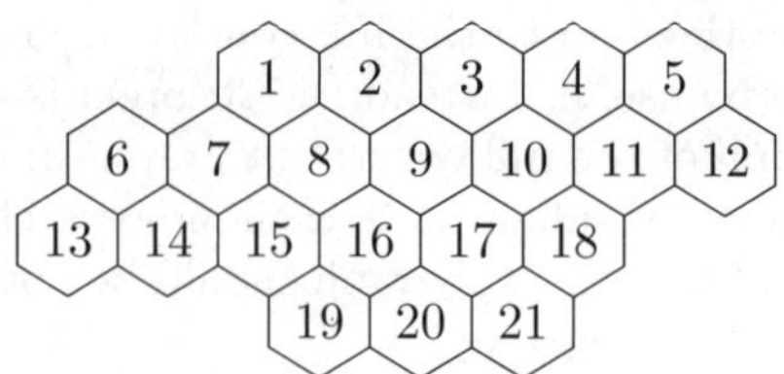

FIG. 4.2.

transmitters in adjacent cells is taken to be 1. Denote by d_k the smallest distance between transmitters which can use frequencies with a separation of k channels. In this example $d_0 = \sqrt{12}$, $d_1 = \sqrt{3}$, $d_2 = d_3 = d_4 = 1$, and $d_5 = 0$, so transmitters within the same cell (co-sited transmitters) must be separated by at least 5 channels.

The subgraph that must be used to obtain the lower bound is too large to apply the software described by Volgenant (1990). Thus, an alternative method is necessary. The form of the travelling salesman bound used here was first obtained by Janssen and Kilakos (1996) using integer programming methods. They showed that if there are only two values of constraint between transmitters in different cells, then it is possible to classify the extreme points of a certain linear programming polytope. One of these extreme points corresponds to the bound described here.

The result is more easily described directly using Theorem 4.8. Let G_s denote the subgraph of the constraint graph corresponding to the transmitters in cell 9 and the six surrounding cells $2, 3, 8, 10, 16, 17$. Consider the graph G'_s, which is a complete graph with no edges of weight 0. It has 275 vertices so $\mathrm{sp}(G_s)$ is at least 274. There is no edge of weight 1 incident with any vertex from cell 9, and hence the channels immediately before and after any channel assigned to one of the 77 vertices in cell 9 must be unoccupied. These unassigned channels are distinct if the co-site constraint is 5 channels. Considering Hamiltonian paths with initial and final vertices in cell 9, we see that $H(G'_s) \geq 274 + 75 \times 2 + 2 \times 1 = 426$. Thus the minimum span for the complete problem is not less than 426. In fact an assignment of span 426 can be found using FASOFT (Hurley *et al.* 1997; Smith *et al.* 1998). A frequency plan of the assignment can be found in Smith *et al.* (1998).[2] Notice that the same bound can be derived (and the same assignment is valid) if the co-site constraint is 3 or 4 channels instead of 5.

4.4 Finding cliques and other subgraphs

The use of subgraphs is generally necessary to calculate good lower bounds. For example, if Theorem 4.8 is applied to a full constraint graph, the number of

[2] The frequency plan can also be seen on the World Wide Web. See Example 3 at: http://www.cs.cf.ac.uk/User/Steve.Hurley/plans.htm.

edges of weight zero is often sufficiently large to allow $H(G')$ to equal zero. A good approach is to find a maximum level-p clique for several values of p, and then choose whichever value of p leads to the best lower bound. A better lower bound is sometimes obtained from a level-p clique with a number of additional vertices. A maximum level-p clique is simply a maximum clique in the subgraph of G obtained by removing all edges with label less than p. The remainder of this section considers how to find maximum cliques in a graph, how the procedure might be simplified for cellular problems and how vertices might be added to improve the bound.

4.4.1 Finding maximum cliques

A maximum clique in a graph is a complete subgraph (not contained in any larger complete subgraph), with the largest possible number of vertices. The problem of determining maximum cliques in a graph is NP-hard. This suggests that, as with the channel assignment problem itself, large problems cannot be solved exactly. However, for the graphs that arise in practice in channel assignment, it appears that the problems can be solved in reasonable time if the number of vertices is less than about 800. For larger graphs it would be necessary to resort to a heuristic procedure, such as that described by Gendreau *et al.* (1993), which would give a large clique, but one not guaranteed to be maximum. In this case the lower bound obtained might be weaker than that given by a true maximum clique.

To find maximum cliques exactly for graphs with fewer than about 800 vertices, the partial enumeration algorithm of Carraghan and Pardalos (1990) gives good results. However, if it is to finish in reasonable time for the larger problems, a good ordering of the vertices should be used. The orderings introduced by Hale (1981) for sequential channel asssignment are effective in this application. Their effectiveness is discussed in more detail by Smith *et al.* (2000).

4.4.2 Maximum cliques in cellular problems

Finding maximum cliques is much less computationally demanding when, as in Example 4.10, there are a small number of transmitter locations and a demand vector which describes the number of transmitters at each site. As the co-site constraint is larger than all other constraints, it is easily seen that if a maximum clique includes one transmitter from a site, it will include all transmitters from the site. This fact can be used (Smith *et al.* 2000) to dramatically reduce the computational requirement of the maximum clique algorithm.

4.4.3 Adding vertices to improve lower bounds

The difficulty of generating lower bounds for practical problems seems to vary considerably. In some cases assignments can be found with spans equal to the clique bound of Theorem 4.6. For others, the best lower bound is obtained from some level-p clique, but a travelling salesman or more sophisticated bound is necessary. The hardest problems are those that require a subgraph larger than a clique. Such problems are not uncommon. It is possible to envisage at least three methods of adding vertices to a clique:

1. Take the chosen level-p clique as the starting subgraph and repeatedly add vertices to the current subgraph as follows. For each vertex not in the subgraph, calculate the sum of the labels of the edges joining the vertex to the current subgraph. Add the vertex for which this sum is largest and re-evaluate the lower bound. If the lower bound does not decrease, continue.
2. Take the chosen level-p clique as the starting subgraph and repeatedly add vertices to the current subgraph as follows. If the current subgraph can be assigned with a span equal to (or close to) the current bound, fix the assignment and attempt to extend it to the full constraint graph without increasing the span. If any vertices are repeatedly involved in constraint violations with the subgraph, they should be added to the subgraph and the bound re-evaluated. If the lower bound does not decrease, continue.
3. Take the chosen level-p clique as the starting subgraph and repeatedly add vertices to the current subgraph as follows. Assign the current subgraph, fix the assignment and calculate the available channels (sometimes called the *reduced frequency domain*) for each vertex not in the subgraph. Add the vertex or vertices with the smallest available number of channels. If the lower bound does not decrease, continue. This is the technique used in FASOFT (Hurley *et al.* 1997).

Although all three methods have their merits and have proved successful on certain problems, each has its problems and none is guaranteed to work. They are also not easy to apply in practice. The development of these methods is a topic which merits further investigation.

4.5 Mathematical programming

There are two reasons for using mathematical programming in lower bounding techniques for channel assignment. First, it may be possible to more easily calculate bounds equivalent to or close to those described in previous sections. For example, the software described by Volgenant (1990) is restricted to a maximum of 250 vertices and does not always converge in a reasonable time. Sometimes the same result can be found using a simple linear program. Second, it is possible to find stronger bounds than the travelling salesman bound using these methods.

The approach of Janssen and Kilakos (1996) using integer programming has already been mentioned. Mathematical programming techniques have also been used in the CALMA project.[3] The emphasis there was on minimizing the number of distinct frequencies used, rather than span minimization.

[3] Information on the CALMA project can be found on the World Wide Web at http://www.win.tue.nl/math/bs/comb_opt/hurkens/calma.html.

4.5.1 Travelling salesman relaxations

The formulation of the travelling salesman problem as an integer program is well known. Let the graph G'_0 be formed from G' by the addition of a dummy vertex v_0 joined by an edge of weight 0 to each vertex of G'. This dummy vertex converts the open symmetric travelling salesman problem to a closed symmetric travelling salesman problem, where it is required to minimize the weight of a circuit instead of the weight of a path. If a minimum weight circuit is found in G'_0, then the vertex v_0 can be removed to give a minimum weight path in G'. Then $H(G')$ is equal to the value of the following integer program for the closed symmetric travelling salesman problem (TSP):

$$\text{Minimize} \quad \sum_{v_iv_j \in E(G'_0)} c_{ij}x_{ij}, \tag{4.1}$$

$$\text{subject to} \quad \sum_{j:v_iv_j \in E(G'_0)} x_{ij} = 2 \quad \text{for all } v_i \in V(G'_0), \tag{4.2}$$

$$\sum_{v_i \in S, v_j \in V(G'_0)\setminus S} x_{ij} \geq 2 \quad \text{for all } S \subset V(G'_0), \tag{4.3}$$

$$x_{ij} \in \{0,1\} \quad \text{for all } v_iv_j \in E(G'_0). \tag{4.4}$$

The formulation gives a minimum weight Hamiltonian circuit in G'_0 from which a minimal weight Hamiltonian path of weight $H(G')$ can be obtained by removing the vertex v_0. The integer variable x_{ij} is equal to 1 if edge v_iv_j is in the Hamiltonian circuit and equal to 0 otherwise. The total weight of the circuit is the sum in (4.1). Equations (4.2) represent the requirement that there are two edges of the Hamiltonian circuit at each vertex. The inequalities (4.3) are the so-called subtour elimination constraints. They ensure that a single circuit is obtained, rather than the union of a number of edge-disjoint circuits.

If this integer program is solved exactly the value of $H(G')$ is obtained. However, this may not be practical. One particular difficulty is the memory required to store the subtour elimination inequalities, as there can be a very large number of them. An approach to making a solution tractable is to relax the problem by weakening or removing one or more of the conditions. If the conditions are weakened it may be possible to find a solution of smaller total weight. Thus $H(G')$ may no longer be obtained, but instead a lower bound for $H(G')$ is found. This lower bound may be still be adequate for the purpose of deriving a strong lower bound for $\text{sp}(G)$.

Replacing the integrality constraint (4.4) by the weaker constraint

$$0 \leq x_{ij} \leq 1 \quad \text{for all } v_iv_j \in E(G'_0) \tag{4.5}$$

gives the linear programming (LP) relaxation of the original integer program. This relaxation can be easy to solve and generally provides a good lower bound. For random graphs the bound is, on average, within 1 per cent of the exact value of $H(G')$. However, the memory requirement problem for the subtour elimination constraints remains.

An alternative is to relax the integer program by removing the subtour elimination constraints (4.3). This gives an integer program for the minimum weight perfect two-matching problem (PTMP) in G'_0. A perfect two-matching is a union of one or more circuits containing every vertex once and once only. An algorithm exists for finding a minimum weight perfect two-matching and guaranteed to terminate in a time which is $O(|V(G)|^2|E(G)|)$ (Pekny and Miller 1994). A lower bound can also be obtained by replacing the integrality constraint (4.4) by (4.5) and using a standard LP solver. When G' corresponds to a clique in some constraint graph, or a clique with some additional vertices, this lower bound is often close to $H(G')$.

The LP relaxation of the minimum weight perfect two-matching problem, applied to a suitable subgraph, appears to be a simple, fast, and robust method of obtaining good lower bounds for $\mathrm{sp}(G)$ in circumstances where the travelling salesman bound is strong. It is, of course, never stronger than the travelling salesman bound.

Example 4.11 As an illustration of the effectiveness of the approach, consider the Philadelphia problem presented in Example 4.10. If we apply the LP relaxation of the minimum weight perfect two-matching problem to the same subgraph as considered previously, generated by cells $2, 3, 8, 9, 10, 16, 17$, then the tight lower bound of 426 is obtained. This approach has the advantage of being applicable to a wider range of problems than the earlier, rather specialized derivation.

The linear programming approach described above is capable of improvement in two circumstances. The first is when the travelling salesman bound is inherently weak, due to the fact that the bound takes no account of constraints between non-consecutive vertices in the Hamiltonian path. Example 4.10 shows that the travelling salesman bound is sometimes tight. It is certainly tight, for example, for a level-p clique where $c_{ij} \leq 2(p+1)$ for all i and j, as then there exist no constraint violations between non-consecutive vertices of the Hamiltonian path. In this case the Hamiltonian path generates an assignment and the bound is best possible. However, there are many problems where the travelling salesman bound is not tight and a stronger bound is necessary. The second source of weakness is where there is some uncertainty about the best subgraph to use for the bound. The travelling salesman bound (and most other bounds) have the property that their value reduces rapidly when more than a small number of critical vertices are added to the appropriate clique. This behaviour can be mitigated if the bound is improved.

4.5.2 Channel assignment constraints

The travelling salesman bound can be improved by adding additional constraints, referred to as channel assignment constraints. Associate a non-negative integer variable e_{ij} with each edge v_iv_j of the constraint graph G. The e_{ij} are chosen so that when an assignment is constructed from a Hamiltonian path $\{v_{i_1},\ldots,v_{i_n}\}$ by setting

$$f(v_{i_1}) = 0$$
$$f(v_{i_j}) = f(v_{i_{j-1}}) + c_{i_{j-1}i_j} + e_{i_{j-1}i_j} \quad \text{for } j = 2,\ldots,n,$$

the assignment will have no constraint violations. Then constraints between consecutive vertices on the Hamiltonian path no longer have to be met exactly, allowing constraints between non-consecutive vertices to be satisfied.

Definition 4.12 *The variable e_{ij} associated with edge v_iv_j is called the excess on the edge v_iv_j.*

To formulate the channel assignment constraints, make the following further definition. If P is a path $v_{i_1}, v_{i_2}, \ldots, v_{i_k}$ with edge set $E(P)$, then let $X_P = x_{i_1i_2} + \cdots + x_{i_{k-1}i_k}$ and $E_P = e_{i_1i_2} + \cdots + e_{i_{k-1}i_k}$.

Definition 4.13 *The* deficit *of the path P, denoted d_P, is defined by*

$$d_P = c_{i_1i_k} - (c_{i_1i_2} + \cdots + c_{i_{k-1}i_k}).$$

Let $\mathcal{P}(G')$ be the set of paths P of G' with $d_P > 0$. Then, if $P \in \mathcal{P}(G')$, it is required that

$$X_P - (|E(P)| - 1) \leq \frac{E_P}{d_P}.$$

To see this, note first that if $X_P \leq (|E(P)|-1)$ then E_P is unconstrained. Second, if $X_P = |E(P)|$ (that is, if all edges of P are included in the Hamiltonian path), then the total excess on P must be at least as large as the deficit of P to ensure that the constraint between the end vertices of P is satisfied. This gives the following integer programming formulation of the channel assignment problem:

$$\text{Minimize} \quad \sum_{v_iv_j \in E(G'_0)} c_{ij}x_{ij} + \sum_{v_iv_j \in E(G')} e_{ij}, \tag{4.6}$$

$$\text{subject to} \quad \sum_{j: v_iv_j \in E(G'_0)} x_{ij} = 2, \qquad v_i \in V(G'_0) \tag{4.7}$$

$$\sum_{v_i \in S, v_j \in V(G'_0)\setminus S} x_{ij} \geq 2, \qquad S \subset V(G'_0) \tag{4.8}$$

$$d_PX_P - d_P(|E(P)| - 1) - E_P \leq 0, \quad P \in \mathcal{P}(G') \tag{4.9}$$

$$x_{ij} \in \{0,1\}, \qquad v_iv_j \in E(G'_0) \tag{4.10}$$

$$e_{ij} \in \{0,1,\ldots,c_{\max}\}, \qquad v_iv_j \in E(G') \tag{4.11}$$

where $c_{\max} = \max_{\{i,j\}} c_{ij}$. Note that when minimizing (4.6), e_{ij} will equal zero if $x_{ij} = 0$.

As with the formulation of the travelling salesman problem, a dummy vertex is used to express the problem in terms of a circuit rather than a path. The objective function (4.6) is the total length of the path to be minimized. Equations (4.7) ensure that there are two edges of the circuit at each vertex. Inequalities (4.8) are the subtour elimination constraints. Inequalities (4.9) ensure that there are no constraint violations between non-consecutive vertices in the path. The fact that the variables x_{ij} are integers is expressed by (4.10) and the fact that the excesses are integers and at most equal to the maximum constraint value is expressed by (4.11).

This integer program, if it can be solved, gives an exact solution of the channel assignment problem for the full constraint graph G. Although this is rarely possible, lower bounds can be derived as follows. Replacing the integrality constraints (4.10) and (4.11) by

$$0 \leq x_{ij} \leq 1, \quad v_i v_j \in E(G'_0) \tag{4.12}$$

$$0 \leq e_{ij} \leq c_{\max}, \quad v_i v_j \in E(G'), \tag{4.13}$$

a linear programming relaxation of the channel assignment problem is obtained. By also omitting the subtour elimination constraints (4.8), a linear programming relaxation of the perfect two-matching problem (PTMP), with extra channel assignment constraints (PTMP + CAP) is obtained. This gives solutions in an acceptable time and appears to give an excellent bound when applied to a suitable subgraph, if it is practicable. Sometimes there are too many constraints in the inequalities (4.9) and a subset of them must be chosen. This can be done by, for example, restricting the length of the paths $P \in \mathcal{P}(G')$ to 3 or 2.

The choice of which channel assignment constraints to keep and which to eliminate may be critical in obtaining a strong bound. However, it appears that elimination of the subtour elimination constraints and relaxation of the integrality constraints makes little or no difference to the strength of the lower bound, for real channel assignment problems. A level-0 clique is often a good candidate subgraph for this method, rather than a higher level clique. This is because the constraints in inequalities (4.9) tend to prevent constraint violations between non-consecutive vertices in the Hamiltonian path, which often occur if a travelling salesman bound is used with this level of clique.

Some results comparing this lower bound with previous methods and illustrating the improvement are given by Allen *et al.* (1999). These results are given for non-cellular problems. In the next section a simplification of this bound for cellular problems will be shown to be capable of generating a tight bound by solving a simple linear program.

4.5.3 A worked example

The formulation PTMP+CAP derived in the previous section has the potential to find the strongest possible bounds. However, the number of channel assignment constraints can, for some problems, be too large for a linear programming package to handle. In this section, a simplification of the method will be described which, for cellular problems, dramatically reduces both the number of variables and the number of constraints. It does this by replacing the variables by certain sums of variables. The method will be illustrated with a recent variation of the Philadelphia problem.

Tcha *et al.* (1997) propose a new lower bound for the span, extending the work of Gamst (1986). They show their bound to be tighter than the Gamst bound and claim wider and easier real-world applicability. The variation of the Philadelphia problem presented is identical to that in Example 4.10 except that cells with centres at distance $\sqrt{3}$ require two channels separation, that is $d_2 = \sqrt{3}$, where previously $d_2 = 1$. Tcha *et al.* find that $\mathrm{sp}(G) \geq 459$. This same value can be obtained using the LP relaxation of the perfect two-matching formulation, applied to the level-0 clique $\{1, 2, 3, 7, 8, 9, 10, 15, 16, 17, 19, 20\}$ with 360 vertices. Notice that the level-0 clique bound of Theorem 4.6 is only 359. The clique is too large to apply Volgenant's software (Volgenant 1990) for the travelling salesman bound. Here it will be shown that a LP relaxation of PTMP + CAP gives a bound $\mathrm{sp}(G) \geq 524$, a result which is best possible.

As before, consider a cellular problem with transmitters located at a number of sites (cells) and with the number of transmitters at each site specified by a demand vector $\boldsymbol{m}$. Let $\mathcal{C}_r$ denote site r, and think of each vertex v in the constraint graph G as being a member of precisely one site. Let $\mathcal{S}_{rs}$ be defined by

$$\mathcal{S}_{rs} = \sum_{i,j} x_{ij}, \quad (v_i \in \mathcal{C}_r,\ v_j \in \mathcal{C}_s)$$

and let $\mathcal{E}_{rs}$ be defined by

$$\mathcal{E}_{rs} = \sum_{i,j} e_{ij}, \quad (v_i \in \mathcal{C}_r,\ v_j \in \mathcal{C}_s).$$

Thus the variable $\mathcal{S}_{rs}$ denotes the sum of the number of edges joining vertices at sites $\mathcal{C}_r$ and $\mathcal{C}_s$ in some Hamiltonian path, and the variable $\mathcal{E}_{rs}$ is the sum of the excesses on these edges.

Let $c = c_{ij}$ $(i, j \in \mathcal{C}_r)$ be a constant co-site constraint and suppose $c > 2c_{ij}$ for all $i \in \mathcal{C}_r$ and $j \in \mathcal{C}_s$ $(r \neq s)$. Then if $\mathcal{S}_{rs} > \min\{m_r, m_s\}$ there are at least $\mathcal{S}_{rs} - \min\{m_r, m_s\}$ edge-disjoint paths of length 2 with the central vertex at one site and the other two vertices at the other site. Any such path requires a non-zero excess on one of the edges. By counting these excesses an inequality of the form

$$(c - 2c_{rs})(\mathcal{S}_{rs} - \min\{m_r, m_s\}) \leq \mathcal{E}_{rs},$$

is obtained. This is a simplified form of the channel assignment constraints for certain paths of length 2.

Again use a dummy vertex v_0, this time considered to be at a dummy site 0 with $m_0 = 1$, in order to express the problem in terms of a circuit. As before, v_0

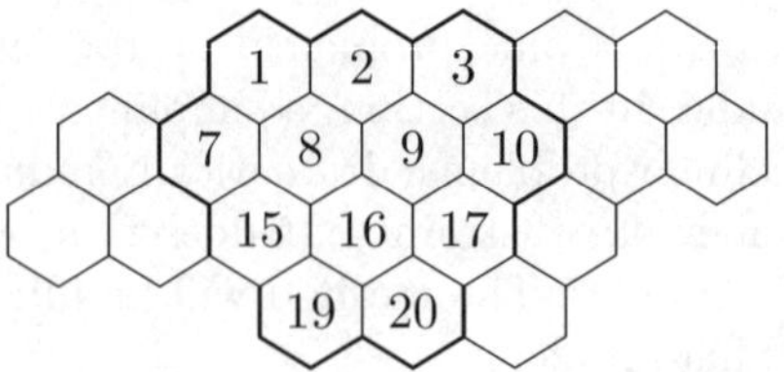

FIG. 4.3.

is joined by an edge of weight 0 to each vertex of G'. The condition that there are two edges at each vertex is expressed as

$$\mathcal{S}_{r0} + 2\mathcal{S}_{rr} + \sum_{s \neq r,\ s \neq 0} \mathcal{S}_{rs} = 2m_r, \quad (r \neq 0)$$

$$\sum_{s \neq 0} \mathcal{S}_{0s} = 2,$$

and the objective function to be minimized is

$$\sum_{r \neq 0} (c\mathcal{S}_{rr} + \mathcal{E}_{rr}) + \sum_{r \neq s,\ r,s \neq 0} (c_{rs}\mathcal{S}_{rs} + \mathcal{E}_{rs})\,, \tag{4.14}$$

where the shorthand c_{rs} is used to denote the constraint between any pair of vertices $v_r \in \mathcal{C}_r$ and $v_s \in \mathcal{C}_s$. The variables are all integer-valued.

Consider the maximum level-0 clique containing the sites shown in Fig. 4.3. The objective function (4.14) is

$$\begin{array}{llllll}
5\mathcal{S}_{1\ 1} & + 5\mathcal{S}_{2\ 2} & + 5\mathcal{S}_{3\ 3} & + 5\mathcal{S}_{7\ 7} & + 5\mathcal{S}_{8\ 8} & + 5\mathcal{S}_{9\ 9} \\
+ 5\mathcal{S}_{10\ 10} & + 5\mathcal{S}_{15\ 15} & + 5\mathcal{S}_{16\ 16} & + 5\mathcal{S}_{17\ 17} & + 5\mathcal{S}_{19\ 19} & + 5\mathcal{S}_{20\ 20} \\
+ 2\mathcal{S}_{1\ 2} & + \mathcal{S}_{1\ 3} & + 2\mathcal{S}_{1\ 7} & + 2\mathcal{S}_{1\ 8} & + 2\mathcal{S}_{1\ 9} & + \mathcal{S}_{1\ 10} \\
+ 2\mathcal{S}_{1\ 15} & + \mathcal{S}_{1\ 16} & + \mathcal{S}_{1\ 17} & + \mathcal{S}_{1\ 19} & + \mathcal{S}_{1\ 20} & + 2\mathcal{S}_{2\ 3} \\
+ 2\mathcal{S}_{2\ 7} & + 2\mathcal{S}_{2\ 8} & + 2\mathcal{S}_{2\ 9} & + 2\mathcal{S}_{2\ 10} & + \mathcal{S}_{2\ 15} & + 2\mathcal{S}_{2\ 16} \\
+ \mathcal{S}_{2\ 17} & + \mathcal{S}_{2\ 19} & + \mathcal{S}_{2\ 20} & + \mathcal{S}_{3\ 7} & + 2\mathcal{S}_{3\ 8} & + 2\mathcal{S}_{3\ 9} \\
+ 2\mathcal{S}_{3\ 10} & + \mathcal{S}_{3\ 15} & + \mathcal{S}_{3\ 16} & + 2\mathcal{S}_{3\ 17} & + \mathcal{S}_{3\ 19} & + \mathcal{S}_{3\ 20} \\
+ 2\mathcal{S}_{7\ 8} & + \mathcal{S}_{7\ 9} & + \mathcal{S}_{7\ 10} & + 2\mathcal{S}_{7\ 15} & + 2\mathcal{S}_{7\ 16} & + \mathcal{S}_{7\ 17} \\
+ \mathcal{S}_{7\ 19} & + \mathcal{S}_{7\ 20} & + 2\mathcal{S}_{8\ 9} & + \mathcal{S}_{8\ 10} & + 2\mathcal{S}_{8\ 15} & + 2\mathcal{S}_{8\ 16} \\
+ 2\mathcal{S}_{8\ 17} & + 2\mathcal{S}_{8\ 19} & + \mathcal{S}_{8\ 20} & + 2\mathcal{S}_{9\ 10} & + 2\mathcal{S}_{9\ 15} & + 2\mathcal{S}_{9\ 16} \\
+ 2\mathcal{S}_{9\ 17} & + \mathcal{S}_{9\ 19} & + 2\mathcal{S}_{9\ 20} & + \mathcal{S}_{10\ 15} & + 2\mathcal{S}_{10\ 16} & + 2\mathcal{S}_{10\ 17} \\
+ \mathcal{S}_{10\ 19} & + \mathcal{S}_{10\ 20} & + 2\mathcal{S}_{15\ 16} & + \mathcal{S}_{15\ 17} & + 2\mathcal{S}_{15\ 19} & + 2\mathcal{S}_{15\ 20} \\
+ 2\mathcal{S}_{16\ 17} & + 2\mathcal{S}_{16\ 19} & + 2\mathcal{S}_{16\ 20} & + 2\mathcal{S}_{17\ 19} & + 2\mathcal{S}_{17\ 20} & + 2\mathcal{S}_{19\ 20} \\
+ \mathcal{E}_{1\ 1} & + \mathcal{E}_{2\ 2} & + \mathcal{E}_{3\ 3} & + \mathcal{E}_{7\ 7} & + \mathcal{E}_{8\ 8} & + \mathcal{E}_{9\ 9} \\
+ \mathcal{E}_{10\ 10} & + \mathcal{E}_{15\ 15} & + \mathcal{E}_{16\ 16} & + \mathcal{E}_{17\ 17} & + \mathcal{S}_{19\ 19} & + \mathcal{E}_{20\ 20} \\
+ \mathcal{E}_{1\ 2} & + \mathcal{E}_{1\ 3} & + \mathcal{E}_{1\ 7} & + \mathcal{E}_{1\ 8} & + \mathcal{E}_{1\ 9} & + \mathcal{E}_{1\ 10} \\
+ \mathcal{E}_{1\ 15} & + \mathcal{E}_{1\ 16} & + \mathcal{E}_{1\ 17} & + \mathcal{E}_{1\ 19} & + \mathcal{E}_{1\ 20} & + \mathcal{E}_{2\ 3} \\
+ \mathcal{E}_{2\ 7} & + \mathcal{E}_{2\ 8} & + \mathcal{E}_{2\ 9} & + \mathcal{E}_{2\ 10} & + \mathcal{E}_{2\ 15} & + \mathcal{E}_{2\ 16} \\
+ \mathcal{E}_{2\ 17} & + \mathcal{E}_{2\ 19} & + \mathcal{E}_{2\ 20} & + \mathcal{E}_{3\ 7} & + \mathcal{E}_{3\ 8} & + \mathcal{E}_{3\ 9}
\end{array}$$

$$
\begin{array}{llllll}
+\mathcal{E}_{3\ 10} & +\ \mathcal{E}_{3\ 15} & +\ \mathcal{E}_{3\ 16} & +\ \mathcal{E}_{3\ 17} & +\ \mathcal{E}_{3\ 19} & +\ \mathcal{E}_{3\ 20} \\
+\mathcal{E}_{7\ 8} & +\ \mathcal{E}_{7\ 9} & +\ \mathcal{E}_{7\ 10} & +\ \mathcal{E}_{7\ 15} & +\ \mathcal{E}_{7\ 16} & +\ \mathcal{E}_{7\ 17} \\
+\mathcal{E}_{7\ 19} & +\ \mathcal{E}_{7\ 20} & +\ \mathcal{E}_{8\ 9} & +\ \mathcal{E}_{8\ 10} & +\ \mathcal{E}_{8\ 15} & +\ \mathcal{E}_{8\ 16} \\
+\mathcal{E}_{8\ 17} & +\ \mathcal{E}_{8\ 19} & +\ \mathcal{E}_{8\ 20} & +\ \mathcal{E}_{9\ 10} & +\ \mathcal{E}_{9\ 15} & +\ \mathcal{E}_{9\ 16} \\
+\mathcal{E}_{9\ 17} & +\ \mathcal{E}_{9\ 19} & +\ \mathcal{E}_{9\ 20} & +\ \mathcal{E}_{10\ 15} & +\ \mathcal{E}_{10\ 16} & +\ \mathcal{E}_{10\ 17} \\
+\mathcal{E}_{10\ 19} & +\ \mathcal{E}_{10\ 20} & +\ \mathcal{E}_{15\ 16} & +\ \mathcal{E}_{15\ 17} & +\ \mathcal{E}_{15\ 19} & +\ \mathcal{E}_{15\ 20} \\
+\mathcal{E}_{16\ 17} & +\ \mathcal{E}_{16\ 19} & +\ \mathcal{E}_{16\ 20} & +\ \mathcal{E}_{17\ 19} & +\ \mathcal{E}_{17\ 20} & +\ \mathcal{E}_{19\ 20}.
\end{array}
$$

The conditions for two edges at each vertex, including the condition for two edges at the dummy vertex, are

$$
\begin{aligned}
&\mathcal{S}_{0\ 1} + 2\mathcal{S}_{1\ 1} + \mathcal{S}_{1\ 2} + \mathcal{S}_{1\ 3} + \mathcal{S}_{1\ 7} + \mathcal{S}_{1\ 8} + \mathcal{S}_{1\ 9} + \mathcal{S}_{1\ 10} \\
&\quad + \mathcal{S}_{1\ 15} + \mathcal{S}_{1\ 16} + \mathcal{S}_{1\ 17} + \mathcal{S}_{1\ 19} + \mathcal{S}_{1\ 20} = 16, \\
&\mathcal{S}_{0\ 2} + \mathcal{S}_{1\ 2} + 2\mathcal{S}_{2\ 2} + \mathcal{S}_{2\ 3} + \mathcal{S}_{2\ 7} + \mathcal{S}_{2\ 8} + \mathcal{S}_{2\ 9} + \mathcal{S}_{2\ 10} \\
&\quad + \mathcal{S}_{2\ 15} + \mathcal{S}_{2\ 16} + \mathcal{S}_{2\ 17} + \mathcal{S}_{2\ 19} + \mathcal{S}_{2\ 20} = 50, \\
&\mathcal{S}_{0\ 3} + \mathcal{S}_{1\ 3} + \mathcal{S}_{2\ 3} + 2\mathcal{S}_{3\ 3} + \mathcal{S}_{3\ 7} + \mathcal{S}_{3\ 8} + \mathcal{S}_{3\ 9} + \mathcal{S}_{3\ 10} \\
&\quad + \mathcal{S}_{3\ 15} + \mathcal{S}_{3\ 16} + \mathcal{S}_{3\ 17} + \mathcal{S}_{3\ 19} + \mathcal{S}_{3\ 20} = 16, \\
&\mathcal{S}_{0\ 7} + \mathcal{S}_{1\ 7} + \mathcal{S}_{2\ 7} + \mathcal{S}_{3\ 7} + 2\mathcal{S}_{7\ 7} + \mathcal{S}_{7\ 8} + \mathcal{S}_{7\ 9} + \mathcal{S}_{7\ 10} \\
&\quad + \mathcal{S}_{7\ 15} + \mathcal{S}_{7\ 16} + \mathcal{S}_{7\ 17} + \mathcal{S}_{7\ 19} + \mathcal{S}_{7\ 20} = 36, \\
&\mathcal{S}_{0\ 8} + \mathcal{S}_{1\ 8} + \mathcal{S}_{2\ 8} + \mathcal{S}_{3\ 8} + \mathcal{S}_{7\ 8} + 2\mathcal{S}_{8\ 8} + \mathcal{S}_{8\ 9} + \mathcal{S}_{8\ 10} \\
&\quad + \mathcal{S}_{8\ 15} + \mathcal{S}_{8\ 16} + \mathcal{S}_{8\ 17} + \mathcal{S}_{8\ 19} + \mathcal{S}_{8\ 20} = 104, \\
&\mathcal{S}_{0\ 9} + \mathcal{S}_{1\ 9} + \mathcal{S}_{2\ 9} + \mathcal{S}_{3\ 9} + \mathcal{S}_{7\ 9} + \mathcal{S}_{8\ 9} + 2\mathcal{S}_{9\ 9} + \mathcal{S}_{9\ 10} \\
&\quad + \mathcal{S}_{9\ 15} + \mathcal{S}_{9\ 16} + \mathcal{S}_{9\ 17} + \mathcal{S}_{9\ 19} + \mathcal{S}_{9\ 20} = 154, \\
&\mathcal{S}_{0\ 10} + \mathcal{S}_{1\ 10} + \mathcal{S}_{2\ 10} + \mathcal{S}_{3\ 10} + \mathcal{S}_{7\ 10} + \mathcal{S}_{8\ 10} + \mathcal{S}_{9\ 10} + 2\mathcal{S}_{10\ 10} \\
&\quad + \mathcal{S}_{10\ 15} + \mathcal{S}_{10\ 16} + \mathcal{S}_{10\ 17} + \mathcal{S}_{10\ 19} + \mathcal{S}_{10\ 20} = 56, \\
&\mathcal{S}_{0\ 15} + \mathcal{S}_{1\ 15} + \mathcal{S}_{2\ 15} + \mathcal{S}_{3\ 15} + \mathcal{S}_{7\ 15} + \mathcal{S}_{8\ 15} + \mathcal{S}_{9\ 15} + \mathcal{S}_{10\ 15} \\
&\quad + 2\mathcal{S}_{15\ 15} + \mathcal{S}_{15\ 16} + \mathcal{S}_{15\ 17} + \mathcal{S}_{15\ 19} + \mathcal{S}_{15\ 20} = 72, \\
&\mathcal{S}_{0\ 16} + \mathcal{S}_{1\ 16} + \mathcal{S}_{2\ 16} + \mathcal{S}_{3\ 16} + \mathcal{S}_{7\ 16} + \mathcal{S}_{8\ 16} + \mathcal{S}_{9\ 16} + \mathcal{S}_{10\ 16} \\
&\quad + \mathcal{S}_{15\ 16} + 2\mathcal{S}_{16\ 16} + \mathcal{S}_{16\ 17} + \mathcal{S}_{16\ 19} + \mathcal{S}_{16\ 20} = 114, \\
&\mathcal{S}_{0\ 17} + \mathcal{S}_{1\ 17} + \mathcal{S}_{2\ 17} + \mathcal{S}_{3\ 17} + \mathcal{S}_{7\ 17} + \mathcal{S}_{8\ 17} + \mathcal{S}_{9\ 17} + \mathcal{S}_{10\ 17} \\
&\quad + \mathcal{S}_{15\ 17} + \mathcal{S}_{16\ 17} + 2\mathcal{S}_{17\ 17} + \mathcal{S}_{17\ 19} + \mathcal{S}_{17\ 20} = 56, \\
&\mathcal{S}_{0\ 19} + \mathcal{S}_{1\ 19} + \mathcal{S}_{2\ 19} + \mathcal{S}_{3\ 19} + \mathcal{S}_{7\ 19} + \mathcal{S}_{8\ 19} + \mathcal{S}_{9\ 19} + \mathcal{S}_{10\ 19} \\
&\quad + \mathcal{S}_{15\ 19} + \mathcal{S}_{16\ 19} + \mathcal{S}_{17\ 19} + 2\mathcal{S}_{19\ 19} + \mathcal{S}_{19\ 20} = 20, \\
&\mathcal{S}_{0\ 20} + \mathcal{S}_{1\ 20} + \mathcal{S}_{2\ 20} + \mathcal{S}_{3\ 20} + \mathcal{S}_{7\ 20} + \mathcal{S}_{8\ 20} + \mathcal{S}_{9\ 20} + \mathcal{S}_{10\ 20} \\
&\quad + \mathcal{S}_{15\ 20} + \mathcal{S}_{16\ 20} + \mathcal{S}_{17\ 20} + \mathcal{S}_{19\ 20} + 2\mathcal{S}_{20\ 20} = 26,
\end{aligned}
$$

$$\mathcal{S}_{0\ 1} + \mathcal{S}_{0\ 2} + \mathcal{S}_{0\ 3} + \mathcal{S}_{0\ 7} + \mathcal{S}_{0\ 8} + \mathcal{S}_{0\ 9} + \mathcal{S}_{0\ 10} + \mathcal{S}_{0\ 15} + \mathcal{S}_{0\ 16} + \mathcal{S}_{0\ 17} + \mathcal{S}_{0\ 19} + \mathcal{S}_{0\ 20} = 2,$$

and a selection of CAP constraints is

$$\begin{aligned}
\mathcal{S}_{8\ 9} - \mathcal{E}_{8\ 9} - 52 &\leq 0,\\
\mathcal{S}_{8\ 16} - \mathcal{E}_{8\ 16} - 52 &\leq 0,\\
\mathcal{S}_{9\ 16} - \mathcal{E}_{9\ 16} - 57 &\leq 0,\\
3\mathcal{S}_{7\ 9} - \mathcal{E}_{7\ 9} - 54 &\leq 0,\\
3\mathcal{S}_{9\ 19} - \mathcal{E}_{9\ 19} - 30 &\leq 0,\\
3\mathcal{S}_{8\ 10} - \mathcal{E}_{8\ 10} - 84 &\leq 0,\\
3\mathcal{S}_{8\ 20} - \mathcal{E}_{8\ 20} - 39 &\leq 0,\\
3\mathcal{S}_{1\ 16} - \mathcal{E}_{1\ 16} - 24 &\leq 0,\\
3\mathcal{S}_{3\ 16} - \mathcal{E}_{3\ 16} - 24 &\leq 0,\\
3\mathcal{S}_{2\ 15} - \mathcal{E}_{2\ 15} - 75 &\leq 0,\\
3\mathcal{S}_{2\ 17} - \mathcal{E}_{2\ 17} - 75 &\leq 0,\\
3\mathcal{S}_{15\ 17} - \mathcal{E}_{15\ 17} - 84 &\leq 0.
\end{aligned}$$

If this linear program is run on a standard LP solver a lower bound of 524 is obtained. This bound is obtained both when it is run as an integer program (with the added condition that all variables be integer-valued) and as a linear program (without such a condition). As an indication of the resources needed, the result is given in 0.24 seconds using CPLEX on a 133 MHz 64 Mb Pentium PC. The same is true if the full set of CAP constraints is generated. In fact the lower bound of 524 is best possible, since the channel assignment package FASOFT (Hurley *et al.* 1997; Smith *et al.* 1998) is able to find an explicit assignment of span 524.

The equations and inequalities appear extensive. However, they can be directly generated from the matrix of elements c_{rs} and the vector $\boldsymbol{m}$ in a form suitable for input to a linear programming package.

The method has proved succesful on other problems. It can, for example, be applied to the original Philadelphia problem described in Example 4.10. If it is applied to the same level-0 clique then the lower bound of 426 is again found. This calculation has the advantage over the original derivation in Section 4.3 that the subgraph is a standard clique, which can be found directly using a maximum clique algorithm.

For cellular problems the method appears to be the best of those described in this chapter. To get a strong bound, the crucial requirement is a suitable subgraph, with a minimum span close to the minimum span of G.

4.6 Bounds for fixed spectrum problems

In the previous sections the constraint graph has been used to find lower bounds on the minimum span of assignments with no interference. In some problems, for example cellular networks, a fixed amount of spectrum is available, which is often insufficient to find an interference-free assignment, that is only assignments with constraint violations are possible. Consequently, it would be useful to have an alternative lower bound, assuming a fixed spectrum is available, and measuring, for example, the minimum number of constraint violations. Work in this area is still in its infancy, but simple lower bounds can be obtained which in some cases may prove useful.

Let q be the given span, so $q + 1$ consecutive channels $1, 2, \ldots, q + 1$ are available for assignment. Given a level-p clique C_p^i, not necessarily maximal, a lower bound on the number of constraint violations in assigning C_p^i using the available span q can be obtained as follows. Let $L(C_p^i) = (p+1)(|V(C_p^i)| - 1)$ be the clique bound of Theorem 4.6. If $q \geq L(C_p^i)$ then there may be an assignment with no constraint violations. If $q < L(C_p^i)$ then the maximum number of vertices that can be assigned without constraint violations is $\lfloor q/(p+1) + 1 \rfloor$. Thus the number of vertices that cannot be assigned in span q is at least

$$|V(C_p^i)| - \left\lfloor \frac{q}{p+1} + 1 \right\rfloor = \frac{L(C_p^i)}{p+1} - \left\lfloor \frac{q}{p+1} \right\rfloor = \left\lceil \frac{L(C_p^i) - q}{p+1} \right\rceil .$$

Each vertex that cannot initially be assigned without creating a constraint violation causes at least one extra constraint violation when it is assigned.

Let $\eta(C_p^i)$ be the lower bound resulting from this argument on the number of constraint violations in assigning the clique C_p^i. Then

$$\eta(C_p^i) = \begin{cases} 0, & \text{if } L(C_p^i) \leq q \\ \left\lceil \dfrac{L(C_p^i) - q}{p+1} \right\rceil, & \text{if } L(C_p^i) > q\,. \end{cases}$$

A lower bound on the number of constraint violations in the constraint graph G can be obtained by summing over all disjoint level-p cliques in G:

$$\eta(G) = \sum_{C_p^i \in G} \eta(C_p^i), \quad (C_p^i \text{ disjoint}).$$

Computationally it may be unrealistic to compute all disjoint cliques C_p^i contained in G; however, in some problem instances, useful, although weaker, lower bounds may be obtained by considering a subset of cliques C_p^i.

Example 4.14 Consider the constraint graph from a radio links problem containing 252 transmitters (Castelino and Stephens 1999). By inspection, the constraint graph contains six level-9 cliques $C_9^i, i = 1, \ldots, 6$ containing 8, 6,

8, 8, 8 and 8 vertices as follows:

$$C_9^1\colon \{1,2,3,4,5,6,7,8\}$$
$$C_9^2\colon \{11,12,13,14,15,16\}$$
$$C_9^3\colon \{17,18,19,20,21,22,23,24\}$$
$$C_9^4\colon \{25,26,27,28,29,30,31,32\}$$
$$C_9^5\colon \{33,34,35,36,37,38,39,40\}$$
$$C_9^6\colon \{41,42,43,44,45,46,47,48\}$$

If 50 channels are available, numbered $1,\ldots,50$ (so $q = 49$), then since $L(C_9^i) = 10(|V(C_9^i)| - 1)$ a lower bound on the number of constraint violations on each level-9 clique gives 3, 1, 3, 3, 3, and 3 violations respectively, that is $\eta(G) = 16$.

An assignment using the simulated annealing fixed spectrum algorithm presented in Hurley *et al.* (1997) gives 20 constraint violations.

4.7 Conclusion

The main techniques for deriving lower bounds for minimum span channel assignment have been surveyed. The techniques based on mathematical programming can now generally give computable and tight bounds for many of the practical problems that arise. For some problems the PTMP bound is already tight. When it is not, the importance of the CAP constraints relative to the subtour elimination and integrality constraints has been noted. The remaining problem is to find general purpose algorithms to generate the best subgraph when a clique is not adequate, or when the constraint graph is too big for the maximum clique algorithm to be applied. In the former case, current methods require substantial manual input and even then are not always satisfactory.

In combination with the best meta-heuristic algorithms (see Chapter 3) these bounds provide good estimates of the necessary span for practical assignments without constraint violations. Such successes depend on the particular structures arising from the geometry of the transmitter locations. If the same techniques are applied to large randomly generated constraint graphs, the gap between the best of these lower bounds and the span of the best assignment known can be very large. It is fortunate that it is the practical problems for which the techniques are most effective.

The situation for fixed spectrum channel assignment is quite different. The study of lower bounds in this case is in its infancy and the discussion given here can be considered as a small initial contribution. Currently, when the available span is significantly less than the minimum span, no accurate estimate of the minimum possible number of constraint violations is available. Thus, the effectiveness of fixed spectrum channel assignment algorithms in these circumstances remains open.

5

CHANNEL ASSIGNMENT ON INFINITE SETS UNDER FREQUENCY–DISTANCE CONSTRAINTS

Jan van den Heuvel and Colin McDiarmid

5.1 Introduction

Large-scale radio systems for multiple users, such as those used in mobile phone networks, are often based on a cellular structure (MacDonald 1979; Hale 1980; Macario 1997). The service area is divided into cells and each cell is serviced by one transmitter, which communicates with users within the cell using a particular radio channel or set of channels. In any particular application the available channels are uniformly spaced in the spectrum, justifying integer labellings of these channels. It is often assumed, at least as a starting position, that the transmitters all have the same power, are omnidirectional, and are laid out like the vertices of a triangular lattice: in this case, the natural cells form a hexagonal pattern.

Suppose that a radio receiver is tuned to a signal on channel c_0, broadcast by the transmitter for its cell (typically, the closest transmitter). Reception will be degraded if there is excessive interference from other transmitters in the vicinity. First, there is 'co-channel' interference due to reuse of channel c_0 at nearby sites; but there are also contributions from sites using channels near c_0, since in practice neither transmitters nor receivers operate exclusively within the frequencies of their assigned channels.

One possible method to decide if an assignment of channels to transmitters is acceptable is as follows. First we agree on a suitable radio propagation model, and a collection of 'test' receiver positions. We then combine the effects of co-channel interference and other interferences at the test receiver positions, and determine whether the corresponding signal to interference ratios are at least some threshold value, at a suitably high proportion of the points. We shall not follow this option: instead we shall adopt the popular and much simpler engineering approach. We assume that, in order to ensure acceptable signal quality, constraints are imposed on the allowed channel separations between pairs of potentially interfering transmitters. Thus, we work within the 'constraint matrix' model. As a further simplification, we assume throughout that the allowed channel separations between a pair of transmitters are uniquely determined by the distance between the transmitters.

In the most basic version of the channel assignment problem, we consider only co-channel interference, and assume that we are given a threshold 'reuse distance' d (or d_0) such that interference will be acceptable as long as no channel is reused at sites less than distance d apart. Given a set V of points in the plane and given $d > 0$, let $G(V, d)$ denote the graph with vertex set V in which distinct vertices u and v are adjacent whenever the Euclidean distance $d(u, v)$ between them is less than d. Our basic version of the channel assignment problem involves colouring such 'proximity' or 'unit disc' graphs. We need to assign colours (radio channels or frequencies) to the points in V (sites or cells or transmitters), using as few colours as possible but avoiding (excessive) interference. A colouring of the vertices of a graph G is called *proper* if adjacent vertices always receive distinct colours. Thus, we are interested in proper colourings of the proximity graph $G(V, d)$ using few colours. The least possible number of colours is the *chromatic number* $\chi(G(V, d))$.

The problem described in the previous paragraph will be considered in Section 5.3. That section is based on McDiarmid and Reed (1999). It turns out that it is possible to make quite precise statements concerning the behaviour of $\chi(G(V, d))$ if either V is a triangular lattice or if the reuse distance d is large. (The phrase 'd is large' really refers to the asymptotic behaviour of $\chi(G(V, d))$ as $d \to \infty$. We need to assume that the set V satisfies a certain mild condition on its density.)

For more practical channel assignment problems we need to consider more than just co-channel interference, and we investigate more general trade-offs between geographical distance and channel separation. Suppose we are given a non-negative vector $\boldsymbol{d} = (d_0, d_1, \ldots, d_{k-1})$ of $k \geq 1$ distances. An assignment (colouring) $\varphi : V \to \{1, 2, \ldots, n\}$ is called $\boldsymbol{d}$-feasible if it satisfies the *frequency–distance constraints*

$$d(u, v) < d_i \quad \Rightarrow \quad |\varphi(u) - \varphi(v)| \neq i,$$

for each pair of distinct points u, v in V and for each $i = 0, 1, \ldots, k-1$. When $d_0 \geq d_1 \geq \cdots \geq d_{k-1}$ we may replace '$\neq i$' above by '$> i$' and we obtain a standard model for channel assignment, see for example Hale (1980). For many practical implementations it can indeed be assumed that $d_0 \geq d_1 \geq \cdots \geq d_{k-1}$, but there exist radio systems that can be modelled better with a description where this is not the case, see for example Section III E in Hale (1980). The *span* $\text{sp}(V; \boldsymbol{d})$ is the least integer n for which there exists a $\boldsymbol{d}$-feasible assignment $\varphi : V \to \{1, 2, \ldots, n\}$. When $k = 1$ we write $\text{sp}(V; d_0)$; note that this is the same as $\chi(G(V, d_0))$.

The values $d_0, d_1, \ldots, d_{k-1}$ are set with the intention that any $\boldsymbol{d}$-feasible assignment will lead to acceptable levels of interference, and it is assumed that the vector $\boldsymbol{d}$ is given. As discussed above, the d_0-constraint limits co-channel interference; and similarly the d_1-constraint limits the contribution to the interference from the 'first adjacent' channels. We wish to develop an understanding of the quantity $\text{sp}(V; \boldsymbol{d})$, and to find methods to construct assignments that achieve the span or close to it. Thus we are interested here in the 'static channel assignment problem with frequency–distance constraints'.

In Section 5.4 we take a look at the behaviour of $\mathrm{sp}(V;\boldsymbol{d})$ for some special cases such as the triangular lattice, and when the distances are large. To consider large distance, we suppose that $\boldsymbol{d} = d\,\boldsymbol{x}$, where $\boldsymbol{x} = (x_0, x_1, \ldots, x_{k-1})$ is a fixed distance k-vector. As in the case when there were co-channel constraints only, we shall again be able to give quite precise information about the behaviour of $\mathrm{sp}(V; d\,\boldsymbol{x})$ for large d. These results are based on the work in McDiarmid (2000). We also consider briefly an extension involving demands and a co-site constraint (Gerke 2000*b*).

In the next section, Section 5.5, we return to the case of co-channel constraints only, and consider another model with demands, where each transmitter may demand a different number of channels. We restrict our discussion to the case of the triangular lattice, which is already quite interesting. This section is based on McDiarmid and Reed (2000).

The results described in Sections 5.3 and 5.4 provide reasonably accurate answers for almost any set V, provided the distances are large enough. In order to obtain more precise answers for small distances, we need to assume more structure on the set V under consideration. In Section 5.6 we discuss some results concerning channel assignments on sets V that are 2-dimensional lattices or have a lattice-like structure. That section is based on Van den Heuvel (1998).

The starting point of most of the research in that section is the result from McDiarmid and Reed (1999) (see also Section 5.3) that assignments that provide a good approximation to $\mathrm{sp}(L; d_0)$ (where L is the set of points of a lattice) can be found among so-called strict tilings (see the definitions in that section). For a constraint vector $\boldsymbol{d} = (d_0, d_1, \ldots, d_{k-1})$ let $\mathrm{sp}_t(L; \boldsymbol{d})$ denote the least integer n for which there exists a $\boldsymbol{d}$-feasible assignment $\varphi : L \to \{1, 2, \ldots, n\}$ which is a strict tiling. Good bounds for $\mathrm{sp}_t(L; d_0)$ follow from the results in Section 5.3. In Section 5.6 we look in some detail at the question of finding those $\boldsymbol{d}$ for which we can guarantee $\mathrm{sp}_t(L; \boldsymbol{d}) = \mathrm{sp}_t(L; d_0)$. In particular it follows that for such $\boldsymbol{d}$ we have good bounds for the span of a feasible assignment.

We thank Stefanie Gerke for several helpful comments.

5.2 Mathematical background and definitions

We will consider points as lying in the Euclidean plane $\mathbf{R}^2$. We will often think of $\mathbf{R}^2$ as a 2-dimensional vector space in the natural way. Thus, we can add and scale points and sets, so that for example $x + A = \{\, x + a : a \in A \,\}$ and $\lambda A = \{\, \lambda a : a \in A \,\}$. A set of the form $x + A$ will be called a *coset* of A.

Also we have the standard inner product $x \cdot y$ and norm $\|x\| = (x \cdot x)^{1/2}$. The (Euclidean) distance between two points is given by $d(x, y) = \|x - y\|$, and the distance between a point x and a set A is given by $d(x, A) = \inf\{\, d(x, a) : a \in A \,\}$.

Some of the definitions in this section were informally given in the previous section: we repeat them here for completeness.

Definition 5.1 *Given a set V of points in the plane and $d > 0$, $G(V,d)$ denotes the graph with vertex set V where two points are adjacent if their Euclidean distance is strictly less than d. The* chromatic number $\chi(G(V,d))$ *is the minimum number of colours needed for a proper colouring of the graph $G(V,d)$.*

Definition 5.2 *Given a set V of points and a positive integer n, an n-*labelling *of V is a function $\varphi : V \to \{1,2,\ldots,n\}$. A* constraint vector *is a non-negative k-vector $\mathbf{d} = (d_0, d_1, \ldots, d_{k-1})$ for some $k \geq 1$. An n-labelling of V is called* $\mathbf{d}$-feasible *if it satisfies the* frequency–distance constraints

$$d(u,v) < d_i \quad \Longrightarrow \quad |\varphi(u) - \varphi(v)| \neq i,$$

for each pair of distinct points u, v in V and for each $i = 0, 1, \ldots, k-1$.

The span $\mathrm{sp}(V;\mathbf{d})$ *is the minimum n such that there exists a $\mathbf{d}$-feasible n-labelling of V.*

Definition 5.3 *A* lattice *L is a subset of the plane of the form $\{\,p\,a + q\,b : p, q \in \mathbf{Z}\,\}$. Here a and b are two given linearly independent vectors that form a* basis *of the lattice. We say that L has as* fundamental cell *the set $F = \{\,x\,a + y\,b : x, y \in \mathbf{R},\ 0 \leq x, y < 1\,\}$.*

A generalized lattice *is a finite union of cosets of a lattice, that is, a set of the form $\cup_{u\in U}(u + L)$, where L is a lattice and U is a finite set of points.*

The triangular lattice T is the lattice with basis $(1,0)$, $(1/2, \sqrt{3}/2)$ (see Fig. 5.1). The square lattice S is the lattice with basis $(1,0)$, $(0,1)$.

Definition 5.4 *A labelling φ of a lattice L is called a* regular tiling *or just* tiling *if there exist linearly independent vectors $s, t \in L$ such that $\varphi(v+s) = \varphi(v+t) = \varphi(v)$ for all $v \in L$. We extend this definition to a labelling φ of a generalized lattice $\cup_{u\in U}(u+L)$ by requiring that $\varphi(v+u+s) = \varphi(v+u+t) = \varphi(v+u)$ for all $v \in L$ and $u \in U$.*

The vectors s, t that define a tiling of a (generalized) lattice form a basis for a sublattice $L^* = \{\,p\,s + q\,t : p, q \in \mathrm{Z}\,\}$ of L. And in fact we can define a tiling as a labelling ϕ such that if $v, w \in L$ with $v - w \in L^*$, then $\varphi(v) = \varphi(w)$.

Definition 5.5 *The sublattice L^* is the* co-channel lattice *of the labelling.*

A strict tiling *of a (generalised) lattice is a tiling that has the additional property that if $v - w \notin L^*$, then $\varphi(v) \neq \varphi(w)$. Hence a strict tiling has the property that all cosets of L^* obtain a different label.*

More special definitions will be given in the text when needed.

5.3 Co-channel constraints

A proximity graph $G(V,d)$ can be considered as a scaled version of a 'unit disc' graphs (Clark *et al.* 1990). When we consider channel assignment problems with only a co-channel constraint, specified by the reuse distance d or d_0, we are interested in proper colourings of such graphs. In Hale (1980) this colouring problem is called the 'frequency–distance constrained co-channel assignment problem'.

We first consider the special case of the triangular lattice graph (the graph originating from transmitter locations with hexagonal cells), which is important in its own right and is the key to understanding the colouring problem in general. We then discuss more general sets such as 'generalized lattices' which still have some structure, and present results that are quite precise and informative, as long as the reuse distance d is not too small. When the reuse distance d is small, small changes in d or in V can lead to large changes in the number of colours needed, and so in order to gain an overview of the problem we consider mainly the case when d is large.

After considering sets as above which have some structure, we consider arbitrary infinite sets of points in the plane, when d is large. It seems reasonable to assume that the points are well spread out. It turns out that we can give good asymptotic results for *any* set V of points in the plane as long as a mild density condition is satisfied. The hope is that such asymptotic results yield insight into finite cases with practical values for the parameters.

Let us then consider first the special case of the triangular lattice, where things work out very neatly. We assume that the lattice has the natural embedding in the plane with minimum distance 1. The origin $O = (0,0)$ and the point $a = (1,0)$ are lattice points, and so are the points $b = (1/2, \sqrt{3}/2)$ and $c = (-1/2, \sqrt{3}/2)$—see Fig. 5.1.

Let G_T denote the corresponding 6-regular graph, with vertex set being the set T of lattice points, and with two vertices adjacent whenever they are at distance 1 in the plane. The six neighbours of the origin O in G_T are then $\pm a, \pm b, \pm c$.

We are interested in colouring the proximity graph $G(T,d)$. For any $d > 0$, let d^+ be the minimum Euclidean distance between two points in T subject to that distance being at least d. Now, the distance between the origin O and the lattice point $p\,a + q\,b$, where p, q are non-negative integers, is $\left((p+q/2)^2 + (\sqrt{3}/2\,q)^2\right)^{1/2} = (p^2+p\,q+q^2)^{1/2}$. Thus, d^+ is the minimum value of

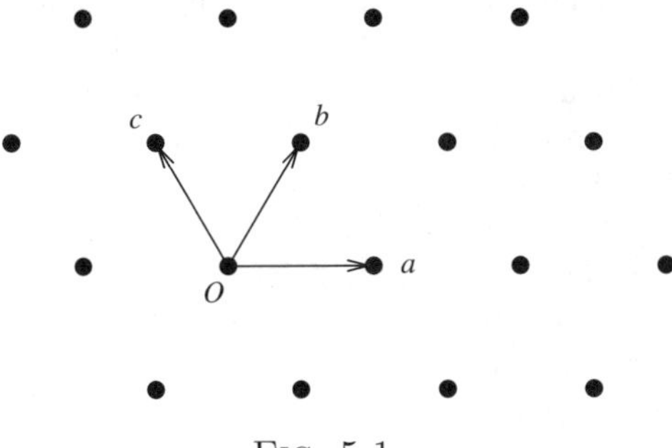

FIG. 5.1.

$(p^2+pq+q^2)^{1/2}$ such that p, q are non-negative integers and $(p^2+pq+q^2)^{1/2} \geq d$. Numbers of the form $p^2 + pq + q^2$ are called *rhombic integers*. The first few rhombic integers are $1, 3, 4, 7, 9, 12, 13, 16, 19, 21$. Note that $d \leq d^+ \leq \lceil d \rceil$, and that we can compute the rhombic integer $(d^+)^2$ quickly, in $O(d)$ arithmetic operations.

Theorem 5.6 *The triangular lattice T satisfies*

$$\chi(G(T, d)) = (d^+)^2$$

for any $d > 0$, and there is an optimal colouring which is a strict tiling, with a triangular co-channel lattice.

This result is proved in McDiarmid and Reed (1999), as indeed are all the results in this section. It appears to have been known to engineers at least since 1979—see MacDonald (1979) and Gamst (1982). A similar result appears as Theorem 3 in Bernstein *et al.* (1997). The above theorem is at the heart of the proof in McDiarmid and Reed (1999) of results for more general sets which we discuss below. Since Theorem 5.6 is such a central result, we include a proof at the end of this section.

It is of interest to note as an aside that we can also work with graph distance in the triangular lattice graph G_T. Given a graph G and positive integer k, let $G^{(k)}$ denote the graph with the same vertices as G, and with distinct vertices u and v adjacent whenever their graph distance in G is at most k. (The graph distance between u and v is the least number of edges in a path joining them.) Thus $G^{(1)}$ is just G. Recall that a *clique* is a set of pairwise adjacent vertices; the *clique number* $\omega(G)$ is the maximum number of vertices in a clique. Clearly $\chi(G) \geq \omega(G)$.

Theorem 5.7 *The graph G_T of the triangular lattice satisfies*

$$\chi\left(G_T^{(k)}\right) = \omega\left(G_T^{(k)}\right) = \lceil 3/4\,(k+1)^2 \rceil$$

for any positive integer k.

This result has been proved independently in Van den Heuvel *et al.* (1998) (where a similar result is given for the graph of the square lattice, with 3/4 replaced by 1/2), as well as in McDiarmid and Reed (1999). It is possible to work simultaneously with Euclidean distance and graph distance: both of the last two theorems are special cases of Theorem 5 of McDiarmid and Reed (1999).

Next, we consider the wider setting of a generalized lattice. Recall that a generalized lattice is a finite union of cosets of a lattice. Observe that these are the only sets which can possibly have a strict tiling. We have just seen that, for the triangular lattice T and any $d > 0$, the proximity graph $G(T, d)$ always has an optimal colouring which is a strict tiling. For any generalized lattice V in the plane, there is a similar approximate result, and thus of course the same holds for any lattice.

Theorem 5.8 *Let V be a generalized lattice in the plane. Then the proximity graph $G(V, d)$ has a colouring which is a strict tiling and which uses at most $\chi(G(V, d)) + O(d)$ colours.*

We shall prove some of the above results concerning lattices and strict tilings at the end of this section.

In order to consider more general sets of points, we shall need to consider densities. Let the lattice L be given as the set of all integer linear combinations of the two linearly independent vectors a and b, with fundamental cell $F = \{x\,a + y\,b : 0 \leq x, y < 1\}$. The sets $\{v + F : v \in L)$ partition the plane, and each contains exactly one point in L. If F has area $A(F)$, it follows that any (large) ball B of radius r contains $\pi\,\sigma(L)\,r^2 + O(r)$ points of L, where $\sigma(L) = 1/A(F)$.

We use this result to guide our definition of density. We say that a set V has *density* $\sigma(V)$ if for any $\epsilon > 0$ there is an r_0 such that for every ball B of radius $r \geq r_0$,

$$\sigma(V) - \epsilon \leq \frac{|V \cap B|}{\pi\,r^2} \leq \sigma(V) + \epsilon.$$

In particular, the lattice L has density $\sigma(L) = 1/A(F)$, and a generalized lattice V which is the union of k distinct cosets of the lattice L has density $\sigma(V) = k\,\sigma(L)$.

Now we consider sets of points that need not be as well structured as generalized lattices, but which are, nevertheless, still approximately uniformly spread over the plane, and which perhaps correspond well to sensible locations for transmitters (extended over the whole plane).

Let $0 < \sigma, r < \infty$. We say that a set V of points in the plane has a *cell structure* with density σ and radius r if there is a family $\{C_v : v \in V\}$ of 'cells' indexed by V such that (a) this family partitions the plane (except perhaps for a set of measure zero); (b) each cell C_v (is measurable and) has area $1/\sigma$; and (c) $C_v \subseteq B(v, r)$ for each $v \in V$. Here $B(v, r)$ denotes the open ball with centre v and radius r. It is easily seen that such a set V has density σ. For example, the set of vertices of the square lattice with unit edge lengths has a cell structure (with square cells) with density 1 and radius $1/\sqrt{2}$. The set T of vertices of the triangular lattice has a cell structure (with hexagonal cells) with density $2/\sqrt{3}$ and radius $1/\sqrt{3}$.

We cannot, now expect to give as precise answers as before. It is of interest to consider other graph parameters related to colouring. The *degree* of a vertex of G is the number of vertices adjacent to it. We denote the maximum degree of a vertex in G by $\Delta(G)$, and the minimum degree by $\delta(G)$.

Theorem 5.9 *Let the set V of points in the plane have a cell structure with density σ and radius r. Then for any $d > 0$,*

$$\pi\,\sigma\,(d - r)^2 - 1 \leq \delta(G(V, d)) \leq \Delta(G(V, d)) \leq \pi\,\sigma\,(d + r)^2 - 1; \qquad (5.1)$$

$$\pi/4\,\sigma\,d^2 \leq \omega(G(V, d)) \leq \pi/4\,\sigma\,(d + 2\,r)^2; \qquad (5.2)$$

and

$$\sqrt{3}/2\,\sigma\,d^2 \leq \chi(G(V,d)) < \left((\sqrt{3}/2\,\sigma)^{1/2}\,(d+2\,r)+2/\sqrt{3}+1\right)^2. \qquad (5.3)$$

Thus both the minimum and the maximum degree of the graph $G(V,d)$ equal $\pi\,\sigma\,d^2+O(d)$, $\omega(G(V,d))=\pi/4\,\sigma\,d^2+O(d)$, and $\chi(G(V,d))=\sqrt{3}/2\,\sigma\,d^2+O(d)$ as $d\to\infty$.

Now we move on to our greatest level of generality. A set of points in the plane is *discrete* if every bounded subset is finite, that is, every ball contains only a finite number of points of the set. Let V be any discrete set of points in the plane. We need a more general notion corresponding to density, which is well defined for any such set V. For each $r>0$ let $f(r)$ be the supremum of the ratio $(|V\cap B|)/(\pi\,r^2)$ over all open balls B of radius r. The *upper density* $\sigma^+(V)$ of V is defined by setting $\sigma^+(V)=\inf_{r>0} f(r)$. If V has a density (e.g., if V is a lattice), then it follows that V has upper density equal to the same value. It turns out (McDiarmid and Reed 1999) that $f(r)\to\sigma^+(V)$ as $r\to\infty$, so that the infimum in the definition is also a limit; and it turns out also that the definition could equally well be phrased in terms of squares rather than balls, or indeed in terms of any 'reasonable' set with finite positive area.

Let us introduce one further graph invariant related to colouring. The *maximin degree* $\delta^*(G)$ is the supremum over all finite induced subgraphs of the minimum degree. This is also called the *degeneracy number* of G, and this number plus 1 is called the *colouring number* of G, since there must be a proper colouring using at most this last number of colours; see for example Jensen and Toft (1995).

Theorem 5.10 *Let V be a discrete set of points in the plane, with upper density $\sigma=\sigma^+(V)$. For any $d>0$, denote the clique number $\omega(G(V,d))$ by ω_d, and use χ_d, Δ_d and δ^*_d similarly for the chromatic number, maximum degree, and maximin degree. Then $\omega_d/d^2\geq\pi/4\,\sigma$ and $\chi_d/d^2\geq\sqrt{3}/2\,\sigma$ for any $d>0$; and, as $d\to\infty$, $\Delta_d/d^2\to\pi\,\sigma$, $\delta^*_d/d^2\to\pi/2\,\sigma$, $\omega_d/d^2\to\pi/4\,\sigma$, and $\chi_d/d^2\to\sqrt{3}/2\,\sigma$.*

It follows, for example, that for any discrete set V of points in the plane with a finite positive upper density, the ratio of the chromatic number of $G(V,d)$ to its clique number tends to $2\sqrt{3}/\pi\sim 1.103$ as $d\to\infty$. It was suggested in Gamst (1986) that such a result should hold for the triangular lattice.

We complete this section on co-channel constraints by giving proofs for the Theorems 5.6 and 5.8 above, where the set V of points has some algebraic structure. For proofs of the other theorems, the reader is referred to McDiarmid and Reed (1999). The upper bounds on the chromatic number in these results come from choosing a suitably scaled version of the triangular lattice to 'cover' the set V, and then transferring a good colouring of the lattice over to V. All the lower bounds on the chromatic number χ come from one idea. Here it is.

Lemma 5.11 *Let V be a discrete set of points in the plane with finite upper density $\sigma = \sigma^+(G)$. Then $\chi(G(V,d)) \geq \sqrt{3}/2\,\sigma\, d^2$ for any $d > 0$.*

Proof A *packing* of disks (or balls) in the plane is a collection of disks with pairwise disjoint interiors. Let $d > 0$ be fixed, and consider disks of diameter d. We shall use Thue's classical theorem (see e.g., Pach and Agarwal (1995), Rogers (1964) and Schneider (1993)), that the maximum density of a packing of equal-sized disks in the plane is achieved by the hexagonal packing. Note that the hexagon with inner radius $d/2$ has area $\sqrt{3}/2\, d^2$. Given a bounded set C, let $\nu(C)$ denote the maximum number of disks with centre in C, over all packings of disks with diameter d. Thue's theorem says that $\nu(B(O,r))/(\pi\, r^2) \to 1/(\sqrt{3}/2\, d^2)$ as $r \to \infty$. Hence, for any $\epsilon > 0$, there is an r_0 such that for any $r \geq r_0$,

$$\nu(B(O,r)) \;\leq\; (1+\epsilon)\,(\pi\, r^2)/(\sqrt{3}/2\, d^2).$$

Now recall that a *stable* (or *independent*) *set* in a graph G is a set of pairwise non-adjacent vertices; and the *stability number* $\alpha(G)$ is the maximum number of vertices in a stable set. Also, the chromatic number $\chi(G)$ is the least number of sets in a partition of the vertex set $V(G)$ into stable sets, and so $\chi(G) \geq |V(G)|/\alpha(G)$. For any ball B_r of radius r,

$$\alpha(G(V \cap B_r, d)) = \nu(V \cap B_r) \;\leq\; \nu(B(O,r)),$$

and so

$$\alpha(G(V \cap B_r, d)) \;\leq\; (1+\epsilon)\,(\pi\, r^2)/(\sqrt{3}/2\, d^2).$$

On the other hand, by the definition of upper density, there is a ball B_r of radius r with $|V \cap B_r| \geq \pi\,\sigma\, r^2$. Hence, for any $r \geq r_0$,

$$\begin{aligned} \chi(G(V,d)) &\geq \frac{|V \cap B_r|}{\alpha(G(V \cap B_r, d))} \\ &\geq \frac{\pi\,\sigma\, r^2}{(1+\epsilon)\,(\pi\, r^2)/(\sqrt{3}/2\, d^2)} \\ &= \frac{\sqrt{3}/2\,\sigma\, d^2}{1+\epsilon}\,. \end{aligned}$$

Since this result holds for any $\epsilon > 0$ we obtain the desired inequality. □

Proof of Theorem 5.6 To establish $(d^+)^2$ as a lower bound on χ, we recall that the triangular lattice with unit edge-lengths has density $\sigma = 2/\sqrt{3}$, and so by Lemma 5.11

$$\chi(G(T,d)) = \chi(G(T,d^+)) \;\geq\; (d^+)^2.$$

We shall show that $(d^+)^2$ is also an upper bound on χ by showing that there is a strict tiling with a triangular co-channel lattice which uses this number of colours.

Recall that the points $a = (1,0)$, $b = (1/2, \sqrt{3}/2)$ and $c = (-1/2, \sqrt{3}/2)$ are neighbours of the origin O in the lattice graph G_T. Recall also that the distance between O and the lattice point $x\,a+y\,b$, where x and y are non-negative integers, is $(x^2 + x\,y + y^2)^{1/2}$.

Now let x_0 and y_0 be non-negative integers such that $(d^+)^2 = x_0^2 + x_0\,y_0 + y_0^2$. Let $p = x_0\,a+y_0\,b$, and $q = x_0\,b+y_0\,c$. Then the set $L = \{\,x\,p+y\,q : x, y \text{ integers}\,\}$ is the set of lattice points of a triangular sublattice of the original triangular lattice. Also, the six points in L closest to the origin O are $\pm p$, $\pm q$, $\pm r$ where $r = x_0\,c-y_0\,a$, and each is at distance d^+. Hence, the minimum distance between distinct points of L is d^+, and so the corresponding strict tiling is a proper colouring of $G(T, d)$.

How many colours does this tiling use? One way to work this out is to note that L is a copy of T scaled by d^+ (and possibly rotated), and so L has density $\sigma(L) = \sigma(T)/(d^+)^2$. Hence, there must be $(d^+)^2$ cosets of L in T. □

Theorem 5.8 will follow easily using the next lemma, from McDiarmid and Reed (1999).

Lemma 5.12 *Let L be a lattice in the plane. Then there are constants c and D such that for any $d \geq D$, there is a sublattice $\hat{L}$ of L which has minimum distance at least d and has density at least $(2/\sqrt{3})/(d^2 + c\,d)$.*

Proof of Theorem 5.8 Let the generalized lattice V consist of k cosets of the lattice L. If V has density $\sigma(V)$ and L has density $\sigma(L)$, then $k = \sigma(V)/\sigma(L)$. By the above lemma, there are constants c and D such that for any $d \geq D$, there is a sublattice $\hat{L}$ of L which has minimum distance at least d and has density $\hat{\sigma} = \sigma(\hat{L})$ at least $(2/\sqrt{3})/(d^2 + c\,d)$. Then V may be partitioned into $(\sigma(V)/\sigma(L)) \cdot (\sigma(L)/\hat{\sigma}) = \sigma(V)/\hat{\sigma}$ cosets of $\hat{L}$. Thus, we have a colouring of V which is a strict tiling and the number of colours used is

$$\sigma(V)/\hat{\sigma} = \sqrt{3}/2\,\sigma(V)\,d^2 + O(d).$$

□

5.4 General frequency–distance constraints

Let us start as in the last section by considering some particular non-asymptotic theorems, which need no lengthy introductions. The first is a simple result for the distance vector $(d, d, \ldots, d)$ and any set V. The second theorem is a general lower bound, which often provides the best known bound, even for particular cases such as the square lattice. The third and fourth theorems are special results for the square and triangular lattices respectively. After these four results, we introduce several asymptotic results, for large distances. All these results are from McDiarmid (2000), though the first result below has been noted also for example in Van den Heuvel (1998).

Recall that, given a set V of points and a constraint k-vector $\boldsymbol{d}$, the *span* $\mathrm{sp}(V; \boldsymbol{d})$ is the minimum n such that there exists a $\boldsymbol{d}$-feasible assignment

$\varphi : V \to \{1, \ldots, n\}$. Thus, when $k = 1$, so that we are considering co-channel constraints only, we have $\mathrm{sp}(V; d) = \chi(G(V, d))$.

Theorem 5.13 *For the k-vector $(d, \ldots, d)$ where $d > 0$,*

$$\mathrm{sp}(V; (d, \ldots, d)) = k \cdot \mathrm{sp}(V; d) - k + 1.$$

Proof Given a d-feasible assignment $\varphi : V \to \{0, 1, \ldots, s-1\}$ where $s = \mathrm{sp}(V; d)$, the assignment $\hat{\varphi}(v) = k \cdot \varphi(v)$ for V is $(d, \ldots, d)$-feasible. Thus $\mathrm{sp}(V; (d, \ldots, d)) \leq k \cdot \mathrm{sp}(V; d) - k + 1$. Conversely, given a $(d, \ldots, d)$-feasible assignment $\varphi : V \to \{0, 1, \ldots, s-1\}$ where $s = \mathrm{sp}(V; (d, \ldots, d))$, the assignment $\tilde{\varphi}(v) = \lfloor \varphi(v)/k \rfloor$ for V is d-feasible (since if u and v are distinct but $\tilde{\varphi}(u) = \tilde{\varphi}(v)$ then $|\varphi(u) - \varphi(v)| < k$, and so $d(u, v) \geq d$). Hence, $\mathrm{sp}(V; d) \leq (\mathrm{sp}(V; (d, \ldots, d)) - 1)/k + 1$. □

Theorem 5.14 *For any set V and any non-increasing distance k-vector $\mathbf{d} = (d_0, \ldots, d_{k-1})$,*

$$\mathrm{sp}(V; \mathbf{d}) \geq \max\{\sqrt{3}/2\, \sigma^+(V)\, (j+1)\, d_j^2 - j : j = 0, \ldots, k-1\}.$$

Proof This theorem follows easily from the above. For any $0 \leq j \leq k - 1$,

$$\begin{aligned} \mathrm{sp}(V; (d_0, d_1, \ldots, d_{k-1})) &\geq \mathrm{sp}(V; (d_0, d_1, \ldots, d_j)) \\ &\geq \mathrm{sp}(V; (d_j, d_j, \ldots, d_j)) \\ &= (j+1)\, \mathrm{sp}(V; d_j) - j. \end{aligned}$$

But

$$\mathrm{sp}(V; d_j) = \chi(G(V, d_j)) \geq \sqrt{3}/2\, \sigma^+(V)\, d_j^2,$$

by Lemma 5.11. □

The following results need more work. In particular, the lower bound in the next theorem should look intuitively plausible, or even obvious, after a little thought, but it takes some time to prove (McDiarmid 2000).

Theorem 5.15 *Let S denote the unit square lattice. Then for any positive integer d,*

$$\mathrm{sp}(S; (\sqrt{2}\, d, d-1)) \leq 2\, d^2,$$

and

$$\mathrm{sp}(S; (\sqrt{2}\, d, d)) \geq 2\, d^2;$$

and so

$$\mathrm{sp}(S; d\, (1, 1/\sqrt{2})) = d^2 + O(d)$$

as $d \to \infty$.

The theorem above is quite precise, but for the triangular lattice we can sometimes give the *exact* span. For any positive integer d, we have

$$\mathrm{sp}(T; (\sqrt{3}\, d, d-1, d-1)) = \mathrm{sp}(T; \sqrt{3}d) = 3d^2.$$

In fact we can extend this result. Recall that a *rhombic integer* is an integer of the form $x^2 + x\,y + y^2$ for (non-negative) integers x and y. These are the squares of the distances between the lattice points of the unit triangular lattice, as we noted earlier; and the first few rhombic integers are $1, 3, 4, 7$.

Theorem 5.16 *Let k be a rhombic integer, let T denote the unit triangular lattice, and let d be a positive integer. Then for the distance k-vector $(k^{1/2}\, d,$ $d-1, \ldots, d-1)$ we have*

$$\mathrm{sp}(T; (k^{1/2}\, d, d-1, \ldots, d-1)) = \mathrm{sp}(T; k^{1/2}\, d) = k\, d^2.$$

Establishing the claimed spans in the two theorems above (and in Theorem 5.23 below) as upper bounds comes from a natural method for constructing assignments, which may be of practical interest. It has the general form: tile the plane, walk through a tile adding k at each step, extend the assignment over the plane and adjust using a k-colouring of the tiles. For example, the following algorithm (where $k = 2$) yields the upper bound in Theorem 5.15 for the unit square lattice (see Fig. 5.2 for the case $d = 3$).

Let d be a positive integer, and consider the sublattice $d\,S$ of the unit square lattice S. Let g be the natural 2-colouring of $d\,S$; that is, $g((idx, jdy))$ is 0 if $i+j$ is even, and is 1 otherwise. The fundamental cell $\{(x, y) : 0 \le x < d,\ 0 \le y < d\}$ for $d\,S$ contains d^2 points of S, forming the set F say. We may trace a path

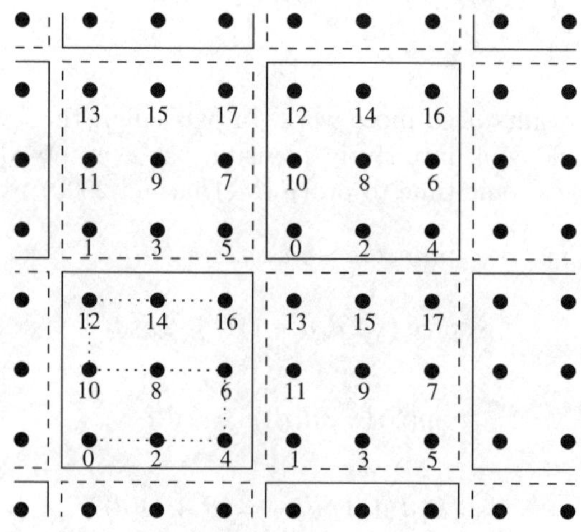

FIG. 5.2.

through the points in F in the lattice graph on S (where points at unit distance are adjacent). Assign the values $0, 2, 4, \ldots, 2d^2 - 2$ to the points along the path, forming the assignment f say on F. The assignments f for F and g for dS yield a natural assignment $h : S \to \{0, 1, \ldots, 2d^2 - 1\}$ for S, as follows. Each point $u \in S$ may be written uniquely as $v + w$ where $v \in F$ and $w \in dS$: we set $h(u) = f(v) + g(w)$. Then h is $(\sqrt{2}\, d, d - 1)$-feasible.

For the asymptotic results on frequency–distance constraints, we need to introduce various densities. We shall restrict our attention here to non-increasing distance vectors. Let $\boldsymbol{x} = (x_0, x_1, \ldots, x_{k-1})$ be a distance k-vector, where $x_0 \geq x_1 \geq \cdots \geq x_{k-1} \geq 0$ and $x_0 > 0$. For each $i = 1, 2, \ldots$, the *i-channel density* $\alpha_i(\boldsymbol{x})$ is defined to be the supremum of the upper density $\sigma^+(V)$ over all sets V of points in the plane for which there is an $\boldsymbol{x}$-feasible assignment using channels $1, \ldots, i$. The 1-channel density $\alpha_1(1)$ is thus the maximum density of a packing of pairwise disjoint unit-diameter discs in the plane; and so $\alpha_1(1) = 2/\sqrt{3}$ and corresponds to taking V as the triangular lattice with unit edge lengths. This is the classical result of Thue on packing discs in the plane, which we met in the proof of Lemma 5.11 above. (We write $\alpha(1)$ instead of $\alpha((1))$ and so on.)

We shall be interested in particular in the 2-channel density $\alpha_2(1, x_1)$. This quantity is the solution of the following red–blue–purple disc packing problem. We wish to pack in the plane a pairwise disjoint family of red unit-diameter discs and a pairwise disjoint family of blue unit-diameter discs, where a red and a blue disc may overlap, forming a purple patch, but their centres must be at least distance x_1 apart. What is the maximum density of such a packing? (Equivalently we may think of packing unit-diameter balls in $\mathbf{R}^3$, where the balls must be in two layers, one with centres on the plane $z = 0$ and one with centres on the plane $z = (1 - x_1^2)^{1/2}$.)

The *channel density* $\alpha(\boldsymbol{x})$ is defined to be the infimum over all positive integers i of $\alpha_i(\boldsymbol{x})/i$. It is not hard to see that $\alpha(1) = \alpha_1(1)$ and so $\alpha(1) = 2/\sqrt{3}$; and that always $0 < \alpha(\boldsymbol{x}) < \infty$. Further, define the *inverse channel density* $\chi(\boldsymbol{x})$ to be $1/\alpha(\boldsymbol{x})$. The next theorem is the central result concerning general frequency–distance constraints when the distances are large.

Theorem 5.17 *For any set V of points in the plane, and any non-increasing distance k-vector $\boldsymbol{x}$*

$$\mathrm{sp}(V; d\,\boldsymbol{x})/d^2 \;\to\; \sigma^+(V)\,\chi(\boldsymbol{x}) \qquad \textit{as } d \to \infty.$$

Thus, in particular, for any set V of points in the plane with upper density 1, such as the set of points of the unit square lattice, the ratio $\mathrm{sp}(V; d\,\boldsymbol{x})/d^2$ tends to the inverse channel density $\chi(\boldsymbol{x})$ as $d \to \infty$.

Following McDiarmid (2000), let us note some basic properties of $\chi(\boldsymbol{x})$. Since $\alpha(1) = 2/\sqrt{3}$ we have $\chi(1) = \sqrt{3}/2$, and so the above result includes the part of Theorem 5.10 above concerning the chromatic number. By rescaling, we can see that for any $a > 0$, $\alpha_i(a\,\boldsymbol{x}) = \alpha_i(\boldsymbol{x})/a^2$ and so $\chi(a\,\boldsymbol{x}) = a^2\,\chi(\boldsymbol{x})$. This is a very useful observation: in particular, we can often set $x_0 = 1$ without loss of generality. Also note that there is an obvious monotonicity, so that if $\boldsymbol{x} \leq \boldsymbol{x}'$, then $\chi(\boldsymbol{x}) \leq \chi(\boldsymbol{x}')$. One further basic property is that the channel density $\alpha(\boldsymbol{x})$ and the inverse channel density $\chi(\boldsymbol{x})$ are continuous functions.

We looked at $\alpha_2(1, x)$ above in different ways. In fact $\chi(1, x) = 2/\alpha_2(1, x)$, which allows us to reduce a problem concerning many channels to one involving just two. Indeed, there is a general result for distance k-vectors $\boldsymbol{x}$ of the form $(1, x, \ldots, x)$.

Theorem 5.18 *Let $0 \leq x \leq 1$ and let $\boldsymbol{x}$ be the k-vector $(1, x, \ldots, x)$. Then*

$$\chi(\boldsymbol{x}) = k/\alpha_k(\boldsymbol{x}).$$

Theorems 5.13–5.16 above yield less precise but tidier results on letting $d \to \infty$, as follows.

Theorem 5.19 *For the k-vector of 1's we have*

$$\chi(1, \ldots, 1) = k \cdot \chi(1) = \sqrt{3}/2\, k.$$

Theorem 5.20 *For any non-increasing distance k-vector $\boldsymbol{x} = (x_0, \ldots, x_{k-1})$,*

$$\chi(\boldsymbol{x}) \geq \sqrt{3}/2 \cdot \max\{\, (j+1)\, x_j^2 : j = 0, \ldots, k-1 \,\}.$$

Theorem 5.21

$$\chi(1, 1/\sqrt{2}) = 1.$$

Theorem 5.22 *Let k be a rhombic integer and let $0 \leq x \leq 1$. Then for the k-vector $(1, x, \ldots, x)$ we have*

$$\chi(1, x, \ldots, x) = \sqrt{3}/2 \cdot \max\{1, k\, x^2\}.$$

Perhaps there is most practical interest in the case $k = 2$, when we have just two distances d_0 and d_1, corresponding to co-channel and first adjacent channel constraints, with $d_0 \geq d_1$. Our current knowledge on the value of $\chi(1, x)$ is summarised in the following theorem. See Fig. 5.3, which shows the lower bounds of the theorem as a solid line and the upper bounds as a broken line.

Theorem 5.23 *We have the exact results from above, that $\chi(1, x) = \sqrt{3}/2$ for $0 \leq x \leq 1/\sqrt{3}$, $\chi(1, 1/\sqrt{2}) = 1$ and $\chi(1, 1) = \sqrt{3}$. We have the lower bounds,*

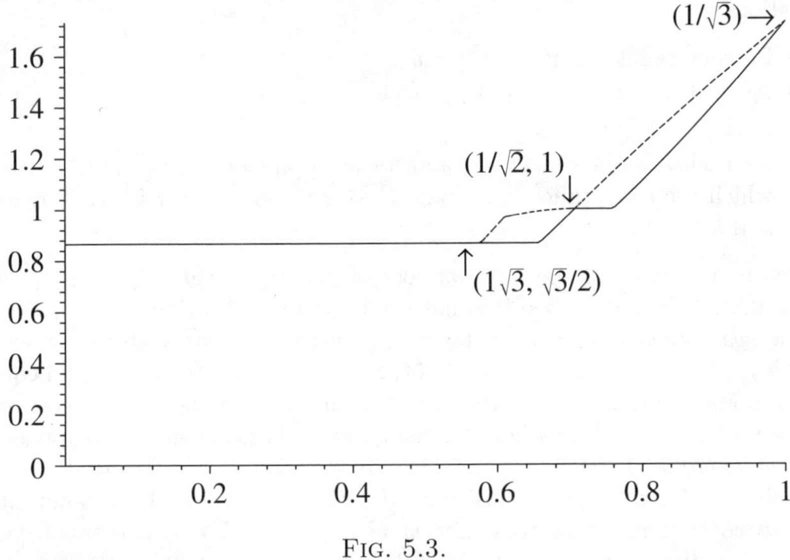

FIG. 5.3.

that $\chi(1,x)$ *is at least*

$$\begin{aligned} \sqrt{3}/2\,, &\quad \textit{for}\ \ 1/\sqrt{3} < x \le 3^{1/4}/2\,(\approx 0.658\,); \\ 2\,x^2\,, &\quad \textit{for}\ 3^{1/4}/2 \le x < 1/\sqrt{2}\ \ (\approx 0.707\,); \\ 1\,, &\quad \textit{for}\ \ 1/\sqrt{2} < x \le 3^{-1/4}\ \ (\approx 0.760\,); \\ \sqrt{3}\,x^2\,, &\quad \textit{for}\ \ 3^{-1/4} \le x < 1. \end{aligned}$$

Finally, we have the upper bounds, that $\chi(1,x)$ *is at most*

$$\begin{aligned} 3\sqrt{3}/2\,x^2\,, &\quad \textit{for}\ \ 1/\sqrt{3} < x \le 4/\sqrt{43}\,(\approx 0.610\,); \\ 2\,x\,(1-x^2)^{1/2}\,, &\quad \textit{for}\ 4/\sqrt{43} \le x < 1/\sqrt{2}\ \ (\approx 0.707\,); \\ 2\,(x^2-1/4)^{1/2}\,, &\quad \textit{for}\ \ 1/\sqrt{2} < x < 1. \end{aligned}$$

Finally in this section, we consider an extension of the central asymptotic result above, Theorem 5.17, which involves demands. Suppose that we are given a discrete set of points V in the plane, and a non-negative integer demand w_v at each point $v \in V$, yielding the *demand vector* $\boldsymbol{w} = (w_v : v \in V)$. As well as a k-vector of distances $\boldsymbol{d} = (d_0, d_1, \ldots, d_{k-1})$ as before, we now also have a positive integer c specifying the co-site constraint. We think of each point v in V as a site to which we are to assign a set of w_v radio channels, subject to frequency–distance and co-site constraints which ensure that interference is not excessive. An assignment $\varphi : V \to \mathcal{P}(\{1, 2, \ldots, t\})$ is called $(\boldsymbol{d}, c)$*-feasible* for $\boldsymbol{w}$ if the following conditions hold on demands, co-site separations and distance separations:

(1) for each point $v \in V$, $|\varphi(v)| = w_v$;
(2) for each point $v \in V$ and for each two distinct elements $a, b \in \varphi(v)$ we have $|a - b| \geq c$;
(3) for each $i = 0, 1, \ldots, k-1$, and for each pair of distinct points $u, v \in V$ which are at distance less than d_i, we have $|a - b| > i$ for each $a \in \varphi(u)$ and $b \in \varphi(v)$.

As before, we may assume without loss of generality that $d_0 \geq d_1 \geq \cdots \geq d_{k-1} > 0$; and since $d_{k-1} > 0$ it is natural to assume also that $c \geq k$.

The *span* $\mathrm{sp}(\boldsymbol{w}; \boldsymbol{d}, c)$ is now the least positive integer t such that there is a $(\boldsymbol{d}, c)$-feasible assignment $\varphi : V \to \mathcal{P}(\{1, 2, \ldots, t\})$ for $\boldsymbol{w}$. If the demand equals 1 at each point, we have exactly the situation discussed earlier in this section. As there we consider the span when the distances are large; that is, we suppose that $\boldsymbol{d} = d\,\boldsymbol{x}$ where $\boldsymbol{x} = (x_0, x_1, \ldots, x_{k-1})$ is fixed and $d \to \infty$. We need to extend the definition for the upper density $\sigma^+(V)$ of a discrete set V of points in the plane to cover demand vectors. Let $\boldsymbol{w} = (w_v : v \in V)$ be a demand vector, where V is a discrete set of points in the plane. For any $r > 0$ let $f(r)$ denote the supremum of the ratio $(\sum_{v \in B} w_v)/(\pi r^2)$ over all open balls B of radius r. The *upper density* of $\boldsymbol{w}$ is defined to be $\sigma^+(\boldsymbol{w}) = \inf_{r>0} f(r)$. The definition could, as with its counterpart $\sigma^+(V)$ for sets, equally well be phrased in terms not of balls but of any reasonable sets with finite positive area. The following extension of Theorem 5.17 above (which leans heavily on that theorem) is due to S. Gerke (private communication).

Theorem 5.24 *Let $\boldsymbol{w} = (w_v : v \in V)$ be a demand vector, where V is a discrete set of points in the plane. Let $\boldsymbol{x} = (x_0, x_1, \ldots, x_{k-1})$ where $x_0 \geq x_1 \geq \cdots \geq x_{k-1} > 0$, and let $c \geq k$. Then*

$$\mathrm{sp}(\boldsymbol{w}; d\,\boldsymbol{x}, c)/d^2 \;\to\; \sigma^+(\boldsymbol{w})\,\chi(\boldsymbol{x}) \quad \text{as } d \to \infty.$$

Proof For each $\delta > 0$, let $\boldsymbol{x}_\delta^+$ denote the c-vector

$$(x_0 + \delta, x_1 + \delta, \ldots, x_{k-1} + \delta, \delta, \ldots, \delta),$$

and let $\boldsymbol{x}_\delta^-$ be the k-vector $\boldsymbol{x} - \delta\,\boldsymbol{1}$, where $\boldsymbol{1}$ denotes the k-vector of 1's.

Let $\epsilon > 0$. By the continuity of the function $\chi(\boldsymbol{x})$ (noted after Theorem 5.17 above), there is a $\delta > 0$ such that $\chi(\boldsymbol{x}_\delta^+) \leq \chi(\boldsymbol{x}) + \epsilon$, $\boldsymbol{x}_\delta^- \geq \boldsymbol{0}$, and $\chi(\boldsymbol{x}_\delta^-) \geq \chi(\boldsymbol{x}) - \epsilon$. Replace each point $v \in V$ by w_v distinct points such that the distance between each new point and v is less than $\delta/2$, and such that all the new points are distinct. We obtain a set V_δ of points in the plane with $\sigma^+(V_\delta) = \sigma^+(\boldsymbol{w})$. We have

$$\mathrm{sp}(\boldsymbol{w}; d\,\boldsymbol{x}, c) \;\leq\; \mathrm{sp}(V_\delta; d\,\boldsymbol{x}_\delta^+)$$

for all $d > 0$, and since $c \geq k$

$$\mathrm{sp}(\boldsymbol{w}; d\,\boldsymbol{x}, c) \geq \mathrm{sp}(V_\delta; d\,\boldsymbol{x}_\delta^-)$$

for all $d \geq 1$. But now Theorem 5.17 above yields

$$\begin{aligned}\limsup_{d\to\infty} \mathrm{sp}(\boldsymbol{w}; d\,\boldsymbol{x}, c)/d^2 &\leq \limsup_{d\to\infty} \mathrm{sp}(V_\delta; d\,\boldsymbol{x}_\delta^+)/d^2 \\ &= \chi(\boldsymbol{x}_\delta^+)\,\sigma^+(V_\delta) \\ &\leq (\chi(\boldsymbol{x}) + \epsilon)\cdot\sigma^+(\boldsymbol{w}),\end{aligned}$$

and

$$\begin{aligned}\liminf_{d\to\infty} \mathrm{sp}(\boldsymbol{w}; d\,\boldsymbol{x}, c)/d^2 &\geq \liminf_{d\to\infty} \mathrm{sp}(V_\delta; d\,\boldsymbol{x}_\delta^-)/d^2 \\ &= \chi(\boldsymbol{x}_\delta^-)\,\sigma^+(V_\delta) \\ &\geq (\chi(\boldsymbol{x}) - \epsilon)\cdot\sigma^+(\boldsymbol{w}).\end{aligned}$$

Since $\epsilon > 0$ was arbitrary, it follows that

$$\begin{aligned}\chi(\boldsymbol{x})\,\sigma^+(\boldsymbol{w}) &\leq \liminf_{d\to\infty} \mathrm{sp}(\boldsymbol{w}; d\,\boldsymbol{x}, c)/d^2 \\ &\leq \limsup_{d\to\infty} \mathrm{sp}(\boldsymbol{w}; d\,\boldsymbol{x}, c)/d^2 \\ &\leq \chi(\boldsymbol{x})\,\sigma^+(\boldsymbol{w}),\end{aligned}$$

which completes the proof. □

5.5 Co-channel constraints with demands

In this section we return to the simple case when there are only co-channel constraints, and when the transmitters are located at vertices of a triangular lattice (with hexagonal cells) in the plane. However, now we allow transmitters to demand more than one channel, as at the end of last section. The number w_v of channels demanded at transmitter v may vary between transmitters. We assume that vertices that are adjacent in the lattice graph must not be assigned the same channel, so as to avoid interference.

The channel assignment problem described above is a 'weighted colouring' problem on the triangular lattice, where the weights correspond to the demands. A *weight vector* for a graph G is a non-zero vector $\boldsymbol{w}$ of non-negative integers w_v indexed by the vertices v of G. Given a graph G and a weight vector $\boldsymbol{w}$, a *weighted colouring* of the pair $(G, \boldsymbol{w})$ is a family of stable (or independent) sets with multiplicities such that each vertex v is in w_v of these sets. The least value of the total number of sets (counting multiplicities) for which there is such a colouring is the *weighted chromatic number* of $(G, \boldsymbol{w})$. A weighted colouring may also be thought of as an assignment to each vertex v of a set of w_v colours such that adjacent vertices receive disjoint sets of colours. The weighted chromatic number is then the least number of colours used in such a colouring. The *weighted colouring problem* is to find a weighted colouring using 'few' colours.

There is a natural graph G_w associated with a pair $(G, \boldsymbol{w})$ as above, obtained by replacing each vertex v by a complete graph on w_v vertices. Weighted colourings of the pair $(G, \boldsymbol{w})$ correspond to usual vertex colourings of the graph G_w, and the weighted chromatic number of $(G, \boldsymbol{w})$ is the chromatic number $\chi(G_w)$ of G_w.

We are interested here in weighted colourings of finite induced subgraphs of the triangular lattice graph, as this corresponds precisely to the basic channel assignment problem described above. We present two theorems from McDiarmid and Reed (2000): the first is a hardness result and the second concerns algorithmic approximation.

Theorem 5.25 *It is NP-complete to determine, on input an induced subgraph G of the triangular lattice graph together with a corresponding weight vector $\boldsymbol{w}$, if the graph G_w is 3-colourable.*

In fact the proof may be easily extended to show that, for any fixed $k \geq 3$, it is NP-complete to tell if G_w is k-colourable. Observe that the graphs G_w here are all unit disc graphs, so this result refines the recent theorem of Gräf *et al.* (1998) that it is NP-complete to tell if a unit disc graph is k-colourable.

Although it is hard to determine $\chi(G_w)$ for a graph G_w as above, it is of course easy to find the maximum size $\omega(G_w)$ of a complete subgraph of G_w in polynomial time. The next result is also from McDiarmid and Reed (2000). Similar results have been put forward in Jordan and Schwabe (1996) and Narayanan and Shende (2001). An extension involving also co-site constraints is given in Schnabel *et al.* (1999). For related work involving co-site constraints see Shepherd (1998) and Gerke (2000*a*).

Theorem 5.26 *There is a polynomial time combinatorial algorithm which, on input an induced subgraph G of the triangular lattice graph together with a corresponding weight vector $\boldsymbol{w}$, finds a weighted colouring of $(G, \boldsymbol{w})$ which uses at most $(4\,\hat{\omega} + 1)/3$ colours, where $\hat{\omega} = \omega(G_w)$.*

The algorithm is quite simple and practical. We first compute the natural 3-colouring f of G, with values 1,2,3; and let $k = \lfloor (\hat{\omega} + 1)/3 \rfloor$. The algorithm then has a distributed phase, in which it constructs $3\,k$ colour sets similar to the colour sets of the colouring f; and finally a tidy-up phase, which involves colouring a weighted forest using new colours. (A forest is a graph in which each component is a tree; it has no cycles and may, therefore, be easily coloured using just two colours.)

In the distributed phase we use the $3\,k$ colours (i, j) for $i = 1, 2, 3$ and $j = 1, \ldots, k$. For each vertex v we compute the value m_v, which is the maximum value of w_x over the neighbours x of v with $f(x) = f(v) + 1 \pmod 3$, and then compute the value $r_v = \min\{w_v - k, k - m_v\}$. We assign the 'low' colours $(f(v), 1), \ldots, (f(v), \min\{k, w_v\})$ to vertex v, and if $r_v > 0$ then we assign also the r_v 'high' colours $(f(v) + 1, k - r_v + 1), \ldots, (f(v) + 1, k)$ borrowed from the vertices coloured $f(v) + 1$. (These colours will not appear on any neighbour of v.) It turns out that the subgraph of G induced by the vertices which have not yet been fully coloured is acyclic, and so the final tidy-up phase is easy.

By using the algorithm, we can find quickly a weighted colouring for an induced subgraph of the triangular lattice such that the number of colours used is no more than about 4/3 times the corresponding clique number of G_w, and hence is no more than about 4/3 times the optimal number. Further by Theorem 5.25 above we cannot expect always to improve on this ratio.

However, perhaps we are being pessimistic. In typical radio channel assignment problems, the maximum number of channels demanded at a transmitter may be quite large. For example, the 'Philadelphia problem' described in Gamst (1986) involves a 21 vertex subgraph of the triangular lattice with demands ranging from 8 to 77 (though it also has constraints on the channels that may be assigned to vertices at distances up to 3, and so it is not a simple weighted colouring problem). Perhaps we can improve on the ratio 4/3 if there are large demands?

We note that the 9-cycle C_9 is an induced subgraph of the triangular lattice graph. Further, for any positive integer k, if we start with a C_9 and replicate each vertex k times, we obtain a graph with clique number $2\,k$ and chromatic number $\lceil 9/4\,k \rceil$. Is this ratio 9/8 of chromatic number to clique number asymptotically worst (greatest) possible? Recent results of Havet (2001) show that if we start with a triangle-free subgraph of the triangular lattice, then we may replace the fraction 4/3 by 7/6. For arbitrary planar triangle-free graphs, the optimal ratio is 3/2. For further discussion on this topic see McDiarmid (2001), Gerke and McDiarmid (2000*a*,*b*).

5.6 Assignments on lattices

In this section we return to problems concerning channel assignment in lattices. Our starting position will be Theorem 5.8 for lattices. So let L be a lattice and d a positive real number. By Theorem 5.8 we know that there exists a feasible labelling of $G(L,d)$ which is a strict tiling and which uses at most $\mathrm{sp}(L;d)+O(d)$ labels. There is a corresponding sublattice L^* of L and a labelling of the cosets of L^* (with all cosets receiving a different label). Notice that this labelling of the cosets is unimportant, apart from the fact that all labels are different, and the only conditions L^* has to satisfy is that it has minimum distance at least d and there are not too many cosets.

Consider now the more general situation in which we have a constraint vector $\boldsymbol{d} = (d_0, d_1, \ldots, d_{k-1})$. Since we know that we can find a strict tiling which gives a good approximation for $\mathrm{sp}(L; d_0)$, an obvious question is if it is possible, by choosing an appropriate labelling of the cosets of the co-channel lattice, to form a $\boldsymbol{d}$-feasible assignment using the same co-channel lattice L^*. A further idea behind this question is that for small values of $d_1, \ldots, d_{k-1}$ the span should be determined by the co-channel reuse distance d_0 only.

Recall that the span $\mathrm{sp}(L; \boldsymbol{d})$ is the minimum n such that there exists a $\boldsymbol{d}$-feasible n-labelling of V. We next define $\mathrm{sp}_t(L; \boldsymbol{d})$ as the minimum n such that there exists a $\boldsymbol{d}$-feasible n-labelling of V which is a strict tiling. This means that $\mathrm{sp}_t(L; \boldsymbol{d}) \geq \mathrm{sp}(L; \boldsymbol{d})$. For examples where $\mathrm{sp}_t(L; \boldsymbol{d}) > \mathrm{sp}(L; \boldsymbol{d})$, see Van den Heuvel *et al.* (1998). In fact, Theorem 5.8 for lattices can be formulated as $\mathrm{sp}_t(L; d) = \mathrm{sp}(L; d) + O(d)$. And the question above can be phrased as:

For which constraint vectors $\boldsymbol{d} = (d_0, d_1, \ldots, d_{k-1})$ can we guarantee $\mathrm{sp}_t(L; \boldsymbol{d}) = \mathrm{sp}_t(L; d_0)$?

In this section we try to give a partial answer to this question. Before we can do so we take a closer look at some of the algebraic properties of lattices and strict tilings of lattices.

Let L be a lattice with basis a, b and let L^* be a sublattice of L, with basis s, t. Hence s, t is a linearly independent pair of vectors from L. As well as considering L as a geometric object, we can consider L as an infinite abelian group (with standard vector addition as the group operation) and L^* as a subgroup of L. Then the quotient group L/L^* is a finite abelian group. If $s = s_1\, a + s_2\, b$ and $t = t_1\, a + t_2\, b$, then it is an easy exercise to deduce that the order of the quotient group is the quotient of the determinants of s, t and a, b; see Lagarias (1995). This means $|L/L^*| = |s_1\, t_2 - s_2\, t_1|$.

Using the observations above, we can view a strict tiling of L with L^* as co-channel lattice as a labelling of the quotient group L/L^* in which each element (each coset of L^*) receives a different label. Recall that a strict tiling of L satisfying the constraint (d_0) with co-channel lattice L^* exists if and only if L^* has minimum distance at least d_0. The number of channels in such a strict tiling is $|L/L^*|$.

The following theorem gives a condition under which it is possible, starting with a strict tiling of L with co-channel lattice L^* having minimum distance at least d_0, to relabel the cosets in order to obtain a strict tiling satisfying the constraint vector $(d_0, \ldots, d_{k-1})$ (and still using $|L/L^*|$ channels).

Recall that if A and B are sets of points, λ a real number and x a point, then $\lambda A = \{\, \lambda a : a \in A \,\}$, $d(x, A) = \inf\{\, d(x, a) : a \in A \,\}$ and $d(A, B) = \inf\{\, d(a, b) : a \in A,\, b \in B \,\}$.

Theorem 5.27 *Let L be a lattice, let $\mathbf{d} = (d_0, \ldots, d_{k-1})$ be a constraint vector, and let L^* be a sublattice of L with minimum distance at least d_0. Suppose that there exists a set $B \subseteq L$ such that the following two conditions are satisfied:*

(1) The set $\{\, b + L^* : b \in B \,\}$ generates the quotient group L/L^*.

(2) $d(i\, C, L^*) \geq d_i$ for each $i = 1, \ldots, k-1$, where C is the convex hull of B.

Then there exists a strict tiling of L satisfying $\mathbf{d}$, using L^* as a co-channel lattice and thus with span $|L/L^*|$.

Some ideas of the proof of Theorem 5.27 will be given at the end of this section.

A drawback of Theorem 5.27 as stated above is that it assumes that a co-channel lattice is given. This is why we next give some consequences of the theorem that make it possible to give more explicit statements on the existence of a strict tiling and on the relation between $\mathrm{sp}_t(L; d_0)$ and $\mathrm{sp}_t(L; \mathbf{d})$. Nevertheless, Theorem 5.27 is useful in its own right. For instance, it can form the basis for an algorithm to find channel assignments on 2-dimensional lattices.

The following is a purely geometrical consequence of the Theorem 5.27. For a lattice L define

$$\gamma_L = \min\{\, \max\{\, \|a\|,\ \|b\|,\ \|a-b\|\,\} : a, b \text{ is a basis of } L\,\}.$$

We will give an alternative definition of γ_L later in this section. For the triangular lattice T we get $\gamma_T = 1$.

Theorem 5.28 *Let L be a lattice, let $\mathbf{d} = (d_0, \ldots, d_{k-1})$ be a constraint vector, and let L^* be a sublattice of L with minimum distance at least d_0. Suppose that there exists a point $z \in \mathbf{R}^2$ such that*

$$d(i\,z,\ L^*) \geq d_i + i\,\gamma_L, \quad \textit{for each } i = 1, \ldots, k-1.$$

Then there exists a strict tiling of L satisfying $\mathbf{d}$, using L^* as a co-channel lattice and thus with span $|L/L^*|$.

Theorem 5.28 makes it possible to derive all kinds of sufficient conditions for the existence of strict tilings, by showing that an appropriate point z exists. In order to do this, without knowing too much about the co-channel lattice L^*, we take a closer look at the geometry of lattices.

Let L be a lattice. A *minimal basis* of L is a pair $m, n \in L$ chosen such that

(1) $\|m\|$ is minimum, subject to $m \in L \setminus \{O\}$;
(2) $\|n\|$ is minimum, subject to $n \in L \setminus \{\, pm : p \in \mathbf{Z}\,\}$;
(3) $m \cdot n \geq 0$.

Note that a minimal basis of a lattice always exists, and is in fact a basis of the lattice. But a minimal basis is not uniquely determined: if m, n is a minimal basis, then so is the pair $-m, -n$. And lattices such as the triangular lattice with many symmetries have even more choices for a minimal basis. Also, it may be shown that if m, n is a minimal basis of a lattice L, then we can easily find the constant γ_L by

$$\gamma_L = \max\{\, \|m\|,\ \|n\|,\ \|m-n\|\,\}.$$

The problem of finding a basis of a lattice satisfying certain minimality conditions, such as those defined above, is a hard problem in high dimensions (see, for example, Section 5.3 of Grötschel *et al.* (1993). But for the 2-dimensional case the so-called *Gaussian Algorithm* provides a fast way to find a minimal basis. For more information on the Gaussian Algorithm, which can be considered as a generalization of the Euclidean Algorithm for finding greatest common divisors, see Daudé *et al.* (1997) and references therein. In this section we need only the following property of a minimal basis, which shows that if a minimal basis is known, then it is straightforward to determine the distance between a given point and the lattice.

Lemma 5.29 *Let x be a point in $\mathbf{R}^2$ and let L be a lattice with minimal basis m, n. Write $x = x_1 m + x_2 n$ and define $x_L = \lfloor x_1 \rfloor m + \lfloor x_2 \rfloor n$ (hence $x_L \in L$). Then*

$$d(x, L) = \min\{\, d(x, x_L),\, d(x, x_L + m),\, d(x, x_L + n),\, d(x, x_L + m + n)\,\}.$$

Now suppose we have a lattice L and a co-channel lattice L^* with minimum distance at least d. Suppose that a, b is a minimal basis of L^*. Then we know that $\|a\| \geq d$ and $\|b\| \geq d$. Let w be the projection of b on the line perpendicular to a, i.e. $w = b - (a \cdot b)\, a/\|a\|^2$, and set $z = 1/2\, a + 1/3\, w$. Figure 5.4 provides a sketch of the situation.

Straightforward planar geometry then shows that $d(z, 0) = d(z, a) \geq 1/\sqrt{3}\, d$, $d(z, b) \geq 1/\sqrt{3}\, d$, and $d(z, a + b) \geq \max\{d(z, 0), d(z, a), d(z, b)\}$. By Lemma 5.29 we can conclude $d(z, L^*) \geq 1/\sqrt{3}\, d$, which together with Theorem 5.28 proves the following result.

Theorem 5.30 *Let d be a positive real number. Then for any lattice L we have $\mathrm{sp}_t(L; (\sqrt{3}\, d, d - \gamma_L)) = \mathrm{sp}_t(L; \sqrt{3}\, d)$.*

By looking somewhat more closely at the arguments used in the proof of this theorem and comparing it with Theorems 5.6 and 5.8, we can show that Theorem 5.30 is in a sense best possible, in particular for the triangular lattice T.

Theorem 5.31 *Let d_0, d_1 be positive real numbers. For any lattice L and for all $\epsilon > 0$, if $d_0 < (\sqrt{3} - \epsilon)\, d_1$ and d_0 is large enough, then $\mathrm{sp}_t(L; (d_0, d_1)) > \mathrm{sp}_t(L; d_0)$. For the particular case of the triangular lattice T, if $\sqrt{3} < d_0^+ \leq \sqrt{3}\, d_1$, then $\mathrm{sp}_t(T; (d_0, d_1)) > \mathrm{sp}_t(T; d_0)$.*

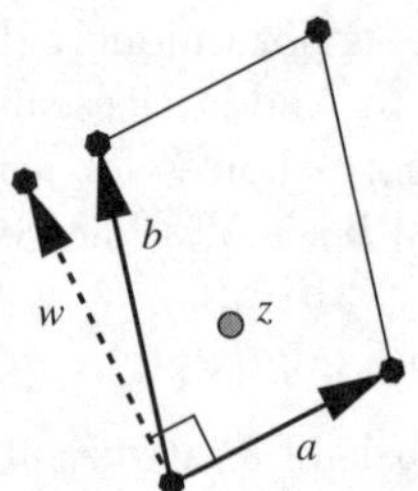

FIG. 5.4.

Bounds for the span of the triangular lattice similar to those in Theorems 5.30 and 5.31 have appeared in Gamst (1982) and Leese (1997).

Theorem 5.30 can be generalized in several directions. No further new ideas are needed to prove the following set of results.

Theorem 5.32 *Let d be a positive real number. Then for any lattice L,*

$$\mathrm{sp}_t(L; (3/\sqrt{2}\,d, d - \gamma_L, d - 2\,\gamma_L)) = \mathrm{sp}_t(L; 3/\sqrt{2}\,d).$$

And for the triangular lattice,

$$\mathrm{sp}_t(T; (\sqrt{3}\,d, d-1, d-2)) = \mathrm{sp}_t(T; \sqrt{3}\,d).$$

Theorem 5.33 *Let $\boldsymbol{d} = (d_0, \ldots, d_{k-1})$ be a constraint vector. For any lattice L, if there exists a point z such that*

$$d(i\,z,\, d_0\,T) \geq \sqrt{6}/2\,(d_i + i\,\gamma_L) \quad \textit{for all } i = 1, \ldots, k-1,$$

then $\mathrm{sp}_t(L; \boldsymbol{d}) = \mathrm{sp}_t(L; d_0)$.

For the triangular lattice T, if there exists a point z such that

$$d(i\,z,\, d_0\,T) \geq d_i + i \quad \textit{for all } i = 1, \ldots, k-1,$$

then $\mathrm{sp}_t(T; \boldsymbol{d}) = \mathrm{sp}_t(T; d_0)$.

The statement about the triangular lattice in Theorem 5.32 is closely related to the formula just after Theorem 5.15, which gives a slightly better result for the case where d is an integer. A similar remark holds for the relation between Theorem 5.16 and the next theorem.

Theorem 5.34 *Let $A > 2^{1/3}\,3^{-1/4}$ (≈ 1.075) and let k be a positive integer and d a positive real number such that*

$$A^2\,k\,d \;\geq\; A\,((1 + \sqrt{3}/2)\,d + 2\,k - 2)\,k^{1/2} + d\,(k-2)$$

(note that this is satisfied for k and d sufficiently large). Then for the distance k-vector $(A\,k^{1/2}\,d, d, \ldots, d)$ we have

$$\mathrm{sp}_t(T; (A\,k^{1/2}\,d, d, \ldots, d)) = \mathrm{sp}_t(T; A\,k^{1/2}\,d).$$

We now give some ideas from the proofs of the results in this section.

Sketch of proof of Theorem 5.27 We use the terminology introduced before the statement of Theorem 5.27. Set $n = |L/L^*|$, so that n is the number of labels used in the strict tiling satisfying (d_0), and let $c_0 = O$ be the origin. Then the statement that there exists a strict tiling of L satisfying $(d_0, \ldots, d_{k-1})$ and using L^* as a co-channel lattice is equivalent to the following claim.

Claim There exists a numbering $c_0 + L^*, c_1 + L^*, \ldots, c_{n-1} + L^*$ of the elements of L/L^* such that if we define the labelling $\varphi : L \to \{1, 2, \ldots, n\}$ of L by $\varphi(v) = i + 1$ if $v \in c_i + L^*$, then this labelling satisfies $(d_0, \ldots, d_{k-1})$.

Since L^* has minimum distance at least d_0, a labelling as defined in the Claim always satisfies (d_0), no matter how the numbering is chosen. Our goal is to show that, under the assumptions of the theorem, the Claim is satisfied.

Now suppose the set $B \subseteq L$ satisfies the conditions in the theorem, in particular the property that $\{ b + L^* : b \in B \}$ generates the group L/L^*. Then it follows easily, using some theory from Cayley digraphs, that we can find a sequence $b_1, b_2, \ldots, b_{n-1}$ of elements in B such that the sequence of partial sum cosets

$$L^*,\ b_1 + L^*,\ b_1 + b_2 + L^*, \ldots, b_1 + b_2 + \cdots + b_{n-1} + L^*$$

are exactly all elements of L/L^*. See Van den Heuvel (1998) for details. Hence the sequence $c_0 = O$, $c_1 = b_1$, $c_2 = b_1 + b_2, \ldots, c_{n-1} = b_1 + b_2 + \cdots + b_{n-1}$ is a numbering of the elements of L/L^* and we will show that this numbering satisfies the Claim.

For all sequences $b_{p+1}, b_{p+2}, \ldots, b_{p+q}$, where $0 \le p < p + q \le n - 1$ we have that

$$c_{p+q} - c_p = b_{p+1} + \cdots + b_{p+q} = q \cdot \left(\frac{1}{q}\, b_{p+1} + \cdots + \frac{1}{q}\, b_{p+q} \right)$$

is an element of $q\,C$ (recall that C is the convex hull of B). Hence by condition (2),

$$d(c_{p+q} - c_p,\ L^*) \ \ge\ d_q.$$

This last statement is equivalent to

$$d(v, w) \ge d_q \quad \text{for all } v \in c_{p+q} + L^* \quad \text{and} \quad w \in c_p + L^*.$$

Since this holds for all p, q with $0 \le p < p+q \le n-1$, the Claim, and hence the Theorem, follows. □

Sketch of proof of Theorem 5.28 Let m, n be a minimal basis of L. Let $D(z, \gamma_L) = \{ x : d(x, z) \le \gamma_L \}$ be the closed ball with centre z and radius γ_L. Using the alternative definition of γ_L above Lemma 5.29, some straightforward geometry shows that there exist $b_1, b_2, b_3 \in L \cap D(z, \gamma_L)$ such that $b_2 - b_1 = m$ and $b_3 - b_1 = n$, or $b_2 - b_1 = -m$ and $b_3 - b_1 = -n$. Set $B = \{b_1, b_2, b_3\}$. It follows that both $m + L^*$ and $n + L^*$ are in the group generated by $\{b_1 + L^*, b_2 + L^*, b_3 + L^*\}$. Since $\{m, n\}$ generates the group L, $\{m + L^*, n + L^*\}$ generates the quotient group L/L^*, hence $\{b_1 + L^*, b_2 + L^*, b_3 + L^*\}$ generates L/L^* as well. This proves that B satisfies condition (1) in Theorem 5.27.

If $a \in D(z, \gamma_L)$, then $d(z, a) \le \gamma_L$ and hence

$$d(i\,a,\ L^*) \ge d(i\,z,\ L^*) - i\,\gamma_L \ge d_i \quad \text{for all } i = 1, \ldots, k - 1.$$

Since the set B lies within $D(z, \gamma_L)$, so does the convex hull of B. So certainly for every vector a in the convex hull of B the inequality above holds, which means that B satisfies condition (2) in Theorem 5.27 as well. □

Recall that a generalised lattice is a set of the form $\cup_{u\in U}(u+L)$, where L is a lattice and U is a finite set of points. Many of the results in this section can be extended to generalized lattices, provided we have some additional information about the set U. For example, the following is an extension of Theorem 5.27 which almost directly follows from the proof of that theorem.

Theorem 5.35 *Let L be a lattice, let $U = \{u_1, u_2, \ldots, u_k\}$ be a finite set of points, and let $\mathbf{d} = (d_0, \ldots, d_{k-1})$ a constraint vector. Let L^* be a sublattice of L with minimum distance at least d_0. Suppose that there exists a set $B \subseteq L$ and a set $V = \{v_2, v_3, \ldots, v_k\}$ satisfying $v_i \in u_i - u_{i-1} + L^*$ for $i = 2, \ldots, k$, such that the following two conditions are satisfied:*

(1) The set $\{b + L^* : b \in B\}$ generates the quotient group L/L^*.
(2) $d(iC, L^*) \geq d_i$ for each $i = 1, \ldots, k-1$, where C denotes the convex hull of $B \cup V$.

Then there exists a strict tiling of $\cup_{u\in U}(u + L)$ satisfying $\mathbf{d}$, using L^* as a co-channel lattice and with span $k\,|L/L^*|$.

6

CONSTRAINT SATISFACTION

Dave Cohen and Peter Jeavons

6.1 Introduction

In this chapter we will introduce the notion of a constraint satisfaction problem (CSP) (Montanari 1974; Dechter and Pearl 1988; Mackworth 1992; Ladkin and Maddux 1994). We will then investigate some of the properties of CSPs, and use the CSP abstraction to represent the frequency assignment problem (FAP) in several different ways.

The main aim of this chapter is to demonstrate how the CSP abstraction can utilize problem *structure* when attempting to solve a problem. To this end we will give examples of the many different approaches to analysing and processing CSPs.

6.2 What is a CSP?

A *CSP* consists of a set of *variables*, each of which is associated with a domain of *values*, and a collection of *constraints*.

Each variable corresponds to a quantity of interest in our problem, for example, the location of a particular transmitter. The domain of values of a variable is the universe from which we can choose possible assignments for that variable. In the case of transmitter locations the domain might be the (finite) set of locations for which we have planning permission!

Each constraint is associated with a set of variables called its *scope*. For the variables in its scope a constraint specifies the simultaneous assignments that are allowed. In this way, constraints represent our knowledge about the interactions between assignments to different variables in the problem.

A solution to a CSP is then an assignment of values to variables of the CSP in such a way that every constraint is satisfied. We are sometimes interested in all solutions to a CSP, sometimes in only one, and sometimes in the decision problem 'does a solution exist?'. Unfortunately, in full generality there is unlikely to be a polynomial solution to any of these problems as the decision problem is NP-complete. It is easy to see this because the classic problem of graph three-colouring can be readily expressed as a CSP. Every node of the graph to be coloured becomes a CSP variable. The domain of each variable is the set of three possible colours. We have a constraint between every pair of variables that correspond to an edge of the original graph three-colouring problem. Each of these constraints allows only different assignments for each of its constrained variables. Since graph three-colouring is NP-complete (Garey and Johnson 1979) it follows that solving the decision problem for CSPs is NP-complete.

Example 6.1 Many scheduling problems can be formulated as CSPs (as described in Section 6.2.2 below). As a simple example we may have four variables $\{p_1, p_2, t_1, t_2\}$. The domain of both p_1 and p_2 is the set {*MachineA*, *MachineB*, *MachineC*}, and the domain of the variables t_1 and t_2 is the positive integers less than 10.

On these variables we have, say, one binary and one four-ary constraint. The binary constraint constrains the pair (t_1, t_2) and states that '$t_1 + 1 < t_2 \vee t_2 + 3 < t_1$'. The four-ary constraint constrains all of the variables and it states that 'if $p_1 = p_2$ then $t_1 \neq t_2$'.

These constraints can have a natural meaning. The binary constraint may mean that one activity takes one minute (or hour or second) and the other takes three minutes, and that they must not overlap. Or perhaps it means that during the first minute of the first activity it would be improper to start the second, and during the first three minutes of the second activity it would be improper to start the first. The four-ary constraint has a more obvious meaning. It is a common rationality constraint which states that 'if two activities happen at the same place then they must happen at different times'!

6.2.1 Representing a CSP

Before we can describe the common CSP representation we need a definition:

Definition 6.2 *(Berge 1973). An (ordered) hypergraph is a pair (V, E) where V is a set of vertices, and E is a set of hyperedges. Each $e \in E$ is an (ordered) subset of vertices in V.*

A hypergraph is labelled if each hyperedge is associated with some label. We often represent a CSP as a labelled ordered hypergraph. Each variable in the CSP is made into a vertex of the hypergraph, and each scope of a constraint is made into an ordered hyperedge. Finally, each hyperedge is labelled with the allowed tuples for the variables in the corresponding constraint.

There are many different but equivalent formalizations of the definition of a CSP, each with advantages and disadvantages. For the purposes of the following discussion we have chosen one standard representation.

Definition 6.3 *A CSP is an ordered five-tuple (V, E, D, σ, ρ) where*

(1) V is a set (of variables),
(2) E is a set of ordered subsets of V (hyperedges or scopes),
(3) D is any set (the union of the domains),
(4) $\sigma : V \Longrightarrow \mathcal{P}(D)$ chooses a domain for each variable, and
(5) $\rho : E \Longrightarrow R$ where R is the set of relations over D.

The image of any edge e of E under ρ must have the same arity as the edge, and is called the constraint relation *over the* constraint scope *e.*

A solution *is an assignment of values to the vertices V in such a way that the ordered list assigned to any hyperedge is a member of the relation labelling that edge.*

A constraint *is a pair $(e, \rho(e))$ where $e \in E$.*

We can normally assume that the domain of all variables is the same, because the assignments to each variable are restricted by the constraints whose scopes it appears in. This implicit selection of domains often makes the representation simpler without any real loss of generality. Sometimes we use a different representation that replaces the function σ with a unary (domain selection) constraint at each variable.

Where necessary, we will give explicit definitions of the functions ρ and σ by naming the domains for variables, and listing the relations associated with hyperedges.

We can now express Example 6.1 in this representation.

Example 6.4 The hypergraph of the problem has four vertices $\{p_1, p_2, t_1, t_2\}$ and only two labelled edges e_1 and e_2. The domain of p_1 and p_2 is the set $\{MachineA, MachineB, MachineC\}$ and the domain of t_1 and t_2 is the positive integers less than 10.

The edge e_1 is the ordered pair (t_1, t_2) and is labelled with the relation $\{(x, y)|x + 1 < y \vee y + 3 < x\}$.

The edge e_2 is the ordered set (p_1, p_2, t_1, t_2) and is labelled with the relation $\{(x, y, z, w)|x \neq y \vee z \neq w\}$.

This example is shown graphically in Fig. 6.1.

The advantage of this representation for us is that we can readily separate out two aspects of the problem:

1. *The roles of the constraint relations*: We may wish to abstract away the interactions of a problem and concentrate on the constraint relations involved.

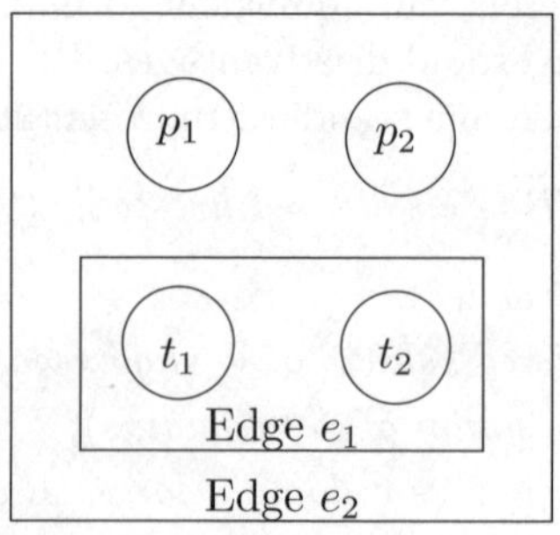

FIG. 6.1.

Each label of a hyperedge is a (constraint) relation of arity given by the size of the corresponding hyperedge (constraint scope). We can consider the *language* for a CSP to be the set of relations occurring in its representation. In this sense, a CSP is a representation of a problem in a particular constraint language. We can then consider whether a particular problem is expressible in another more restrictive (or simpler) CSP language.

As there are many different representations for the same problem, a CSP may be expressible in many different languages.

Techniques exist which are extremely efficient for certain constraint languages. So this is a powerful abstraction (see Section 6.4.2).

2. *The ways in which their scopes intersect*: In particular we can isolate the way in which different variables interact by choosing to ignore the labels (constraint relations) on the hyperedges of the representation. This abstraction is rich in that it allows us to use the mature field of Graph Theory to analyse the structure of a CSP without considering the particular constraint relations involved. Many of the practical techniques for pre-processing CSPs come from an analysis of their hypergraph structure (see Section 6.4.1).

6.2.2 Example: machine shop scheduling is a CSP

We have a machine shop containing various machines. We have a job that will require us to use the various machines for the various tasks involved. This is a very general framework and can be expressed naturally as a CSP.

One way to do this is to construct three CSP variables for each task to be performed. Two of these have time as their domain and they represent the start time and the end time of the task. The domain of the third variable is the (finite) set of possible machines and it represents the location for this task.

To model the problem we use some binary constraints (constraints whose scope is a pair of variables) and some six-ary constraints (whose scope is six variables). To model the fact that one task must finish before another can begin we use a binary constraint between the end time of the first and the start time of the second. To model the fact that if two processes happen to be scheduled on the same machine they cannot overlap in time we use a six-ary constraint. The scope of this constraint is the two location variables and all four relevant time variables. The constraint relation specifies that, when the two tasks are at the same location, then one of them must entirely precede the other.

6.2.3 Example: frequency assignment is a CSP

There are many different forms of the frequency assignment problem (FAP). In this example we give just one way to model a particular form of the FAP with a CSP.

In this case we have a required coverage area for each base station. These may correspond to the areas of coverage for local TV or radio stations. For each of these coverage areas we are to build a single servicing transmitter.

We may choose to model each transmitter using three variables, representing its location, power, and frequency. The variables are then constrained to give sufficient coverage for each base station, given an appropriate interference model.

The coverage condition for each transmitter can be expressed as a binary constraint between the position and the power of the transmitter. The further a transmitter is from the centre of its coverage area the stronger its signal must be.

If we assume a binary interference model (i.e. assuming a single dominant interferer) then we obtain interference conditions relating the frequency, location, and strength of each pair of transmitters. These are complicated six-ary constraints.

An alternative is to use a more realistic interference model and restrict configurations of nearby transmitters to those that we know will not give inappropriate interference. This gives different arities of constraint on different groups of transmitters.

6.2.4 Summary

CSPs are a very general framework for the representation of problems. Effective use of the CSP paradigm involves two stages. The problem has to be effectively presented as a constraint problem, and then the constraint problem must be effectively solved.

In the next section we consider how to alter the formulation of a CSP in the hope of finding a more effective representation. This will lead us onto the subsequent section where we will consider bounds on the complexity of solving a CSP based on its relational and hypergraph structure.

6.3 Representational issues

In this section we will consider the modelling problem. That is, the method of representing a problem as a CSP. We have to begin by choosing just one CSP representation. As with all modelling, this requires a mixture of experience and ingenuity and is best learnt by practice. However, when we have a CSP it is always possible to change its representation in several ways before attempting a solution. It is some of the possible techniques for modifying an existing CSP representation, before solving it, that we will consider here.

6.3.1 Pre-processing techniques

Some of the techniques that can be applied to CSPs preserve the original problem variables and the original set of solutions. These are best thought of as pre-processing techniques.

The most useful of the pre-processing techniques are those which guarantee levels of *consistency* (Mackworth 1977; Freuder 1985; Dechter and van Beek 1997).

Definition 6.5 *A (partial)* assignment *to the variables of a CSP is a (partial) function from the variables to their domains. A partial assignment is valid if it satisfies all constraints whose scopes have been entirely assigned. A CSP is called*

```
INPUT: A CSP
OUTPUT: A k-consistent CSP
---------------------------------------------------------
done = FALSE
WHILE NOT done
   done = TRUE
   FOR every (k − 1)-tuple of variables V
      FOR every variable x not in V
         FOR every valid assignment l of V
            IF l does not extend to V ∪ {x} THEN
               RESTRICT the constraint on V by excluding l.
               done = FALSE
            ENDIF
         ENDFOR
      ENDFOR
   ENDFOR
ENDWHILE
```

Fig. 6.2.

k-consistent if every valid partial assignment to $k-1$ variables can be extended to a valid partial assignment including any other variable.

Given an arbitrary CSP we can make it k-consistent for some k. The way this has to be done is by introducing (possibly new) $(k-1)$-ary constraints. The purpose of these new constraints is to eliminate all assignments to sets of $k-1$ variables that cannot be extended.

A naive algorithm to guarantee k-consistency is given in Fig. 6.2.

The RESTRICT statement means that if there is already a constraint c whose scope is exactly V, then simply remove the labelling l from the relation of c. If there does not exist a constraint whose scope is equal to V then one is created that allows every labelling except l.

There are N choose $(k-1)$ possible $(k-1)$ tuples of variables in a CSP with N variables. Each of these $(k-1)$ tuples allows at most d^{k-1} different labellings, where d is the size of the domain D. Since every loop removes one labelling from one of these $k-1$ tuples the algorithm must terminate in at most N choose $k-1$ times d^{k-1} iterations. Each iteration involves at most $N-k+1$ steps. Applying naive bounds, the algorithm terminates in $O(N^k d^{k-1})$ steps and so is polynomial of degree k. Optimal algorithms for achieving k-consistency are given in Cooper (1989).

6.3.1.1 Different levels of consistency Naturally, to achieve higher levels of consistency requires more work. As consistency does not alter the set of solutions there is a trade-off between work done in achieving a level of consistency and work done during the subsequent solution algorithm.

Two-consistency is relatively cheap to achieve, and is called *arc consistency*. There are many very efficient algorithms available for achieving arc consistency (van Hentenryck *et al.* 1992; Schiex *et al.* 1996). Three-consistency is called path consistency and is slightly more expensive to achieve. Less work has been done on path consistency, but still there are algorithms available which are much more efficient than the naive algorithm given in this section (Tsang 1993).

Strong k-consistency is j-consistency for every $j \leq k$. Strong k-consistency can also be obtained in a tractable fashion. When a CSP has been made strong k-consistent it implies that any backtracking algorithm will assign at least k variables before it requires to backtrack. What is hoped is that achieving some finite level of strong consistency will be sufficient to guarantee that a backtracking algorithm (see Section 6.4.1) will always complete without the need to backtrack. What is almost sure is that ensuring any level of consistency will reduce the amount of backtracking required.

6.3.1.2 Backtrack freeness It is natural to ask at this point, whether a CSP can be solved effectively by ensuring a sufficient level of consistency.

We say that a CSP is backtrack free if the standard backtracking algorithm (see Section 6.4.1) will never have cause to execute a backtrack step. Backtrack freeness is equivalent to the property of being strong N-consistent for a problem with N variables. A backtrack-free CSP is essentially solved.

There is a well known result which states that backtrack freeness is achievable in a binary CSP with a tree structure by ensuring arc consistency (Freuder 1982). In general there is no $k < N$ for a problem with N variables, such that ensuring k-consistency will guarantee backtrack freeness.

However, for a given problem, there is a level of (strong) consistency, which, if it already pertains in the problem, will guarantee backtrack freeness.

Theorem 6.6 (*Dechter 1992*). *Let P be a constraint satisfaction problem with domain size d and let r be the length of the largest scope in P.*

If P is strong $d(r-1)+1$ consistent then it is globally consistent.

This level of consistency is dependent only on the (maximum) domain size and the size of the largest scope. The practical problem with this result is that, in order to achieve a certain level of consistency, new constraints are generated. The sufficiency condition requires consistency of a higher degree than the size of the largest scope. If the consistency algorithm adds any constraints, then the size of the largest constraint has been increased. In this case, we then need to ensure a higher level of consistency. Once again, this may induce new constraints. Hence, for a given problem we cannot *a priori* restrict the level of consistency required to ensure backtrack freeness.

However, if a CSP already exhibits the (sufficient) required level of consistency then we can conclude that it will be solved in a backtrack free manner.

6.3.2 *Alternative formulations*

In this section we will give some more general techniques for changing the representation of a CSP. We try to ensure that solutions to the changed representation are encodings of solutions to the original representation.

Naively, this can be done by reducing the problem to one variable whose domain encodes the solutions to the original problem. It can be seen that this does not help to identify problem structure, but rather obfuscates it. The major reason to change problem formulation is to elucidate problem structure with a view to finding efficient solution techniques.

There are still several approaches. One is to abstract away certain problem details in order to obtain a simpler problem. Solutions to the simpler problem can then be refined into solutions to the original problem. This is the approach often used in the FAP. However it is unlikely to produce solutions that are optimal, as important problem details may have been excluded.

In fact, the FAP is often modelled as a binary problem (using a single dominant interferer model). One of the major advantages of the general constraint model is that it allows higher order interactions to be taken into account (recall Section 6.2.3).

In the next few sections we consider certain standard alternative formulations for CSPs which do not lose any information.

6.3.2.1 Variable value swapping One simple approach to reformulating a problem as a new CSP is to exchange the roles of values and variables. In the new CSP we will be trying to decide which sets of variable are assigned to a particular domain value, rather than which domain value each variable is assigned. Although this is a straightforward concept the mechanics are quite complicated.

We start with a constraint problem $P = (V, E, D, \sigma, \rho)$, and we will construct the equivalent variable/value swapped CSP $P' = (V', E', D', \sigma', \rho')$.

The set of variables V' of the new problem is the union of the domains of the original problem. That is, each new variable will be named with a value from one of the old domains. For each new variable we now need to choose its domain. The domain of each of the new variables is the power set of the old variables. So, choosing a value for a variable in the new CSP corresponds to choosing which set of variables of the old CSP should be labelled with a particular (old) domain value. Formally, $D' = \mathcal{P}(V)$, and σ' is the constant function whose image is D'.

Lastly we must choose the constraints for the reformulation. We first need some definitions.

Definition 6.7 *Given a CSP $P = (V, E, D, \sigma, \rho)$ and a partial assignment l of the variables of P.*

We say that l is a bad partial assignment *if*

(1) The variables over which l is defined form an edge $e \in E$,

(2) l does not satisfy $\rho(e)$.

A subset of D is possibly bad *if it is the set of values of a bad partial assignment of P.*

Let B be the set of bad partial assignments of P.

Choose any ordering of D and with respect to this order let E' be the set of ordered possibly bad subsets of D. For any $b = (b_1, \dots, b_k) \in E'$, let

$$\rho'(b) = \left\{(x_1, \dots, x_k) \in D'^k | \nexists f \in B, (\forall i, \{v | f(v) = b_i\}) \subseteq x_i)\right\}.$$

This is a very complicated definition and is best explained with reference to an example. Once again we refer to the example of the FAP.

Example 6.8 We first give a simple CSP for representing the FAP. The variables represent transmitters and the domain for all variables is the set of allowed frequencies. The constraints restrict subsets of the transmitters and prevent them being assigned frequencies that could cause interference.

Assume that we have four transmitters, each of which needs a frequency. The frequencies for transmitters t_1 and t_2 must differ by more than one channel. Transmitters t_2, t_3, and t_4 cannot all use the same frequency, and transmitters t_1, t_3 and t_4 cannot all use the same frequency.

We begin with the CSP $F_1 = (V = \{t_1, t_2, t_3, t_4\}, E = \{e_1, e_2, e_3\}, D = \{1, 2, \dots, 10\}, \sigma, \rho)$ where

$$\begin{aligned} e_1 &= (t_1, t_2), \\ e_2 &= (t_2, t_3, t_4), \\ e_3 &= (t_1, t_3, t_4), \end{aligned}$$

σ is the constant function whose image is the whole of D, and

$$\begin{aligned} \rho(e_1) &= \{(x, y) | x - y > 1 \lor y - x > 1\}, \\ \rho(e_2) &= \{(x, y, z) | x \neq y \lor y \neq z\}, \\ \rho(e_3) &= \{(x, y, z) | x \neq y \lor y \neq z\}. \end{aligned}$$

Using the variable/value swapping, the FAP, F_1, can be expressed as a CSP where the variables are the frequencies to be assigned and their values are the sets of transmitters given that frequency.

After variable/value swapping we obtain

$$F_2 = (V', E', D', \sigma', \rho'),$$

where

$$\begin{aligned} V' &= D, \\ D' &= \mathcal{P}(V), \\ \forall v \in V', \quad \sigma'(v) &= D'. \end{aligned}$$

In other words, we now have ten variables, corresponding to the ten possible frequencies of the original problem. A solution to F_2 will be an assignment to each of these frequencies of a set of transmitters that will use that frequency.

The bad partial assignments of F_1 are

$$\begin{array}{lll} \text{For } i = 2, \ldots, 10 & l_i^+ : e_1 \to D, & l_i^+(t_1) = i - 1,\ l_i^+(t_2) = i \\ \text{For } i = 1, \ldots, 10 & l_i^= : e_1 \to D, & l_i^=(t_1) = i,\ l_i^=(t_2) = i \\ \text{For } i = 1, \ldots, 9 & l_i^- : e_1 \to D, & l_i^-(t_1) = i + 1,\ l_i^-(t_2) = i \\ \text{For } i = 1, \ldots, 10 & l_i^2 : e_2 \to D, & l_i^2(t_2) = i,\ l_i^2(t_3) = i,\ l_i^2(t_4) = i \\ \text{For } i = 1, \ldots, 10 & l_i^3 : e_3 \to D, & l_i^3(t_1) = i,\ l_i^3(t_3) = i,\ l_i^3(t_4) = i \end{array}$$

The possibly bad subsets of D are then the sets of values of these 48 functions. Some of these functions (e.g. $l_7^=$ and l_7^3) have the same sets of values (in this case the set $\{7\}$). In fact there are only 19 different possibly bad sets. We choose to order the values of D in the natural way and we get the 19 edges of E':

$$\begin{array}{lc} \text{For } i = 1, \ldots, 9 & (i, i+1). \\ \text{For } i = 1, \ldots, 10 & (i). \end{array}$$

We now construct ρ'.

Take for example the new edge (4). There are three bad partial assignments, $l_4^=, l_4^2$, and l_4^3 whose values are the set $\{4\}$. Following the definition we must allow all tuples for this edge except those which correspond to one of these partial assignments. We get that

$$\rho'((4)) = \{(x) \in D' \mid (\{t_1, t_2\} \not\subseteq x) \wedge (\{t_2, t_3, t_4\} \not\subseteq x) \wedge (\{t_1, t_3, t_4\} \not\subseteq x)\}.$$

In fact, for all ten hyperedges e' with just one vertex we get the relation

$$\rho'(e') = \{(x) \mid \nexists e \in E, e \subseteq x\},$$

and for the other nine hyperedges e' with two vertices we get

$$\rho'(e') = \{(x, y) \mid (t_1 \notin x \vee t_2 \notin y) \wedge (t_1 \notin y \vee t_2 \notin x)\}.$$

The meaning of F_2 can be understood as follows. It allows us to assign any set of transmitters to each frequency, subject to only two restrictions. The first is that we cannot have sets of transmitters containing all three of t_2, t_3, and t_4, or all three of t_1, t_3, and t_4, or both of t_1 and t_2, (from the hyperedges of F_2 with one vertex) assigned to the same frequency. The second restriction is that we also cannot assign sets of transmitters containing t_1 and t_2 respectively to adjacent frequencies (from the hyperedges of F_2 with two vertices).

As it stands F_2 allows all the valid solutions of F_1. However, it also allows some solutions which were not available in F_1. In F_2 there is no restriction which prevents a transmitter from being assigned more than one frequency. F_2 makes sure that, no matter how many frequencies each transmitter gets assigned, the interference conditions of F_1 are met. It is possible to extend F_2 by adding additional constraints which limit the number of frequencies assigned to each transmitter, and indeed assure that each transmitter is assigned at least one frequency! It is left as an exercise for the reader to do this extension.

Aside from this nice property of allowing different sorts of solutions after encoding, what does value/variable swapping give us? The main result of value variable swapping in the FAP is that the original domain was *essentially uniform*. That is, the different frequencies in the domain of the transmitters all have approximately the same properties. This was not true of the transmitters. Each transmitter is in a unique relationship to all the others which is captured in the hypergraph structure of F_1. The uniformity of the domain of F_1 is transformed into a symmetry among the variables of F_2. This allows us to apply a partial solution anywhere to any subset of the variables. In fact, it allows us to build up a good exemplary set of partial solutions and use some optimization technique to apply them to the variables of F_2. So, a particular aspect of the problem structure (the uniformity of the domain) is brought to the fore in F_2 and suggests novel ways of attacking the FAP. Indeed this approach has been used and has given good results (Leese 1996).

6.3.2.2 Dual problems In this section we will describe a common technique for transforming any CSP into a binary problem. The method essentially encodes all of the information involved in any constraint into the structure of the domain of a new variable. It comes in two flavours—dualization and redundant variables.

Before considering full dualization we will introduce the notion of a redundant variable used to encode a constraint.

6.3.2.3 Encoding constraints as domain structure Given any CSP we choose any constraint c to be encoded. We intend to remove the constraint c and to replace it with only binary constraints. What makes this result surprising is that it seems that some information is *essentially* non-binary.

Consider the constraint c on e_2 of the CSP F_1 in Example 6.8. We will try to replace c by three binary constraints on the three pairs of variables $(t_1, t_2), (t_2, t_3)$, and (t_1, t_3). If we are not to restrict the allowed assignments to e_2 then we must allow every assignment to, for instance, (t_1, t_2) that c allows. But, c allows *every* assignment to (t_1, t_2)! Similarly c allows any assignment to each of the other two pairs of variables. This implies that all three binary constraints must allow every assignment. However, then the three together allow every assignment to e_2. Hence we have shown that there are some constraints which are essentially non-binary.

So, could we do better by allowing the luxury of new (redundant) variables. These are variables whose values are unimportant in the consideration of a solution, but are used to constrain variables of interest. We continue the example to demonstrate how this is done.

First, enumerate all of the tuples in the constraint relation of c. In this case there are 990 of them, that is the 1000 possibilities minus the ten excluded tuples. We now add a new variable x whose domain has 990 values, each corresponding to a tuple of the relation of c. We now add three binary constraints, which connect x to each of the three variables in the scope of c. The constraint relation of each new binary constraint only allows one assignment for each value in the domain of x. Consider the constraint between t_1 and x. As each domain value of x corresponds to an assignment to e_2 it implies a 'corresponding' assignment to t_1. Each labelling in the relation of this constraint only allows the domain values of x with their corresponding assignment to t_1. Similarly for the two other new binary constraints.

We now remove the constraint c. Clearly, the new configuration of four variables with three constraints allows exactly 990 assignments, one per domain value of x. It should also be clear that these assignments allow exactly the assignments to e_2 as did the original ternary constraint. We have encoded the constraint complexity into the domain of a new redundant variable.

If we encode each constraint of a CSP in this way then we can reformulate any CSP as a binary CSP.

The next method is very similar in intent but different in realization.

6.3.2.4 Dualization For this method we go one step further. As well as removing all of the original constraints we remove all of the original variables!

The method is simple. Again we enumerate the assignments of each of the constraint relations. Once again we create new variables, one for each constraint, whose domains are the encodings of the original assignments. The difference comes in that we now create new binary constraints among the new variables.

For every pair of new variables we consider their old scopes. If these intersect then we add a binary constraint between them. This binary constraint allows only those tuples for the new variables that correspond to assignments that were equal on the intersection of the old scopes. After doing this, we remove the old variables (and their constraints) completely. The resulting CSP is called the *dual* of the original CSP.

In order to show the equivalence of the two formulations we show how valid assignments to each of the CSPs correspond.

First, consider a valid assignment to the original CSP. This is an assignment of values to variables such that the assignment to the variables of the scope of any constraint is contained within the constraint relation. So a valid assignment picks out a labelling from each constraint relation in such a way that, whenever scopes intersect, the same values are chosen. This is a valid assignment for the dual.

Suppose, instead, that we have a valid assignment for the dual CSP. This corresponds to choosing a labelling from each of the constraint relations of the original CSP. Since the constraints in the dual CSP make sure that variables which were in the scopes of two original constraints are assigned the same value it follows that we can recover a consistent assignment of values to variables in the original CSP. Of course, this assignment is valid in the original CSP as, by definition of the dual, the assignment to variables in the scope of a constraint are in the relation of that constraint.

So we have shown that the two representations are entirely equivalent as far as mapping solutions from one to the other is concerned.

Example 6.9 Recall the FAP presented in Example 6.8.

We will now construct the dual of this FAP $F_3 = (\hat{V}, \hat{E}, \hat{D}, \hat{\sigma}, \hat{\rho})$.

In the dual we set $\hat{V} = E$.

We create a constraint in the dual exactly when the two original constraints constrain a common variable. In this case we need three (binary) constraints, one between each pair of variables

$$\hat{E} = \{(e_1, e_2), (e_2, e_3), (e_1, e_3)\}.$$

The domain of e_1 is the 72 allowed pairs for e_1 in F_1, that is $\hat{\sigma}(e_1) = \{(x, y) \in D^2 | x - y > 1 \vee y - x > 1\}$.

The domains of e_2 and of e_3 are equal and each contains 990 values, that is $\hat{\sigma}(e_2) = \hat{\sigma}(e_3) = \{(x, y, z) \in D^3 | x \neq y \vee y \neq z\}$.

The constraints make sure that the common variables of F_1 should have the same values.

For the edge (e_1, e_2) there is only one common variable, namely t_2. So the relation labelling this edge is given by

$$\hat{\rho}((e_1, e_2)) = \{((x, y), (y, z, w)) | (x, y) \in \hat{\sigma}(e_1) \wedge (y, z, w) \in \hat{\sigma}(e_2)\}.$$

For the edge (e_1, e_3) there is only one common variable, namely t_1. So the relation labelling this edge is given by

$$\hat{\rho}((e_1, e_3)) = \{((x, y), (x, z, w)) | (x, y) \in \hat{\sigma}(e_1) \wedge (x, z, w) \in \hat{\sigma}(e_3)\}.$$

For the edge (e_1, e_2) there are two common variables, namely t_3 and t_4. So the relation labelling this edge is given by

$$\hat{\rho}((e_2, e_3)) = \{((x, y, z), (w, y, z)) | (x, y, z) \in \hat{\sigma}(e_2) \wedge (w, y, z) \in \hat{\sigma}(e_3)\}.$$

We know, from the argument above, that the CSP F_3 encodes exactly the same information as F_1. However, it is interesting to compare the different presentations. In F_1 we have four variables with domain size 10, and three constraints with scopes of two, three, and three variables. These constraints have relations with 72, 990, and 990 labellings respectively.

In the dual problem, F_3, we have three variables and three binary constraints. The domains of the three variables have 72, 990, and 990 values, and the constraint relations have 7128, 9810, and 9810 labellings.

6.3.3 Summary

We can increase the consistency of a CSP to make solving it easier (and sometimes backtrack free). We can swap the roles of values and variables. We can add redundant variables, or we can fully dualize a CSP. These methods each explore the internal structure of a problem before attempting to solve it. In many cases we employ more than one of these techniques when solving a problem.

6.4 Properties of a CSP

Once we have formulated our problem as a CSP we then have to solve the CSP. Much research has been done into the way in which the underlying structure of a CSP can be used to determine the complexity of its solution. The complexity is usually bounded by appealing to properties of the underlying constraint hypergraph, or of the constraint language. We will consider these two cases separately.

6.4.1 Graph structure

As mentioned at the end of Section 6.2.1, if we abstract away the relations from a CSP we are left with the underlying hypergraph structure. We then use the hypergraph structure to bound the amount of work that would need to be done to solve the CSP. To do this, we assume a standard backtracking algorithm that assigns values to variables. We can then use the hypergraph structure to bound the maximum depth of backtracking.

More specifically, we are going to measure the computational complexity of solving a particular CSP in terms of the number of assignments to variables made during backtracking when searching for the first solution.

The standard backtracking algorithm, given in Fig. 6.3, assigns values in the domain of variables sequentially. After each assignment to a variable all of the constraints involving the currently assigned variables are checked. If any are violated then a different value for that variable is chosen. If we are successful in assigning a domain value for a variable, then a new variable is chosen and the algorithm continues. If no domain value is possible, then the algorithm *backtracks* to the previously assigned variable and tries a new value for that variable. This algorithm is *complete* in that, if there is any solution it will find one, and it is *correct* in that every solution it generates satisfies all of the constraints.

While standard backtracking can be improved in many ways it is still a good benchmark for assessing the underlying complexity of a CSP.

Without any further information, the complexity of backtracking for solving a CSP with N variables and domain size d is $O(d^N)$.

```
INPUT: A CSP
OUTPUT: A single solution or NONE
--------------------------------------------------------------
SORT the set of variables V as v_1, v_2, ..., v_n

/* assign is a partial function from V into their domains */
/* untried is a partial function from V to subsets of their domains */

assign := null function
untried := null function
index := 1
WHILE index > 0 ∧ index < n
   possible := ∅
   ; X will be a subset of the domain of v_index
   X := {all valid extensions of assign to v_index}
   IF X is NOT empty THEN
      extend untried by setting untried(v_index) == X
   ELSE
      index := index - 1
      WHILE index ≠ 0 ∧ untried(v_index) == ∅
         /* Each iteration through this loop is called a backtrack step */
         index := index - 1
      ENDWHILE
   ENDIF
   choose an element p of untried(v_index)
   modify untried by removing p from the image of v_index
   extend assign by setting assign(v_index) = p
ENDWHILE
IF index > 0 THEN
   return assign
ELSE
   return NONE
ENDIF
```

FIG. 6.3.

Most of the results concerning the hypergraph structure of a CSP concern the case when all constraints are binary and the constraint hypergraph is a standard graph.

6.4.1.1 Graph decomposition Clearly, if we could solve subsets of the constraints independently then we have reduced the complexity of the problem. In fact, if we decompose the original problem with N variables into two problems, one with N_1 variables, and the other with N_2 variables, then we have reduced the original complexity of $O(d^N)$ to a complexity of $O(d^{N_1} + d^{N_2})$ which is a dramatic improvement.

We can obviously bound the backtracking in this way if the graph is disconnected since we can then clearly solve each connected component independently!

Carrying on from this observation, we call a point an *articulation point* if its removal disconnects a graph. If a graph contains an articulation point then we solve the subproblem consisting of variables on one side of the articulation point. Then we use these solutions to reduce the domain of the variable labelling the articulation point. Then we continue by solving the variables on the other side with this reduced domain for the common variable. So, the presence of an articulation point reduces the complexity of a CSP in much the same way as a disconnected component. If the two disconnected components after removing the articulation point have N_1 and N_2 variables respectively, then the complexity of solving is reduced to $O(d^{N_1+1} + d^{N_2+1})$.

6.4.1.2 Tree (acyclic) structure The following theorem bounds the complexity of a (binary) constraint problem if the underlying graph structure is a tree. In this case, we can reduce the general exponential complexity to a polynomial bound.

Theorem 6.10 *(Freuder 1982; Montanari 1974). Let C_{tree} be the class of all binary constraint satisfaction problems for which the associated constraint graph is a tree. There exists a polynomial time algorithm which solves any constraint in C_{tree}.*

In fact this result is the graph specialization of a more general hypergraph result which applies to non-binary CSPs. Before we can discuss this more general result we need some technical definitions.

Definition 6.11 *An edge of a hypergraph is called a* separating edge *if, after removing all of the points contained in that edge the remaining hypergraph is disconnected. The components with respect to such an edge are called* hinges.

A hinge is a minimal hinge *if it cannot be decomposed into smaller hinges.*

This definition is important as it has been shown that hinge structure is the key to using decomposition to reduce the complexity of a CSP hypergraph (Gyssens *et al.* 1994). The first result concerning hinges is that every hypergraph yields a decomposition into a *hinge tree.*

Definition 6.12 *A hinge tree for a hypergraph H is a tree T of nodes N_i where*

(1) Every node N_i in T is a hinge of H,

(2) Whenever N_i and N_j are connected in T, the common vertices of N_i and N_j (i.e. $\cup N_i \cap \cup N_j$) *are a separating edge of H,*

(3) Every edge of H appears in at least one node of T.

A minimal hinge tree *is a hinge tree all of whose nodes are minimal hinges.*

The number and size of the hinges in different minimal hinge trees for the same hypergraph can vary but the size of the largest hinge (in terms of the number of edges in the hinge) in any such decomposition is an invariant of the hypergraph (Gyssens *et al.* 1994). The invariant K is called the *degree of cyclicity* of a hypergraph.

Using this result, we can bound the complexity of solving a general CSP, as follows. We first solve each of the hinges as an independent CSP, and then combine them along the tree structure, using Theorem 6.10 to bound the complexity of the process.

If the number of edges in the largest hinge in the hinge tree is K then we can solve a CSP with N variables and E edges, whose largest constraint has size l in $O(NE^2) + O(El^K K \log l)$ steps. Another bound referring to the domain size, d, is $O(NE^2) + O(Ed^K K \log d)$ which is again a clear improvement on $O(d^N)$ when K is small.

Hinge decompositions properly refine decompositions using articulation points which properly refine decompositions into separate connected components. Furthermore, hinge decompositions extend the notion of a tree structure to general hypergraphs. The graph concept of being tree structured has the natural analogy of being an acyclic hypergraph. Being an acyclic hypergraph is fully characterized by having degree of cyclicity 2.

There is one further result on hinges (Gyssens *et al.* 1994) which guarantees that hinges are the most refined decomposition that is possible without consideration of the relational structure of a CSP. This result states that, if H is a subproblem of a CSP with underlying hypergraph G, and if every (valid) assignment to H can be extended to a solution, then H is a hinge of G. The theorem implies that hinges are the smallest structures that a CSP can be decomposed into effectively.

6.4.1.3 Other graph approaches There are two further approaches which yield good complexity bounds. These are the *cycle cutset* method and the *tree clustering* method (Dechter and Pearl 1988). The cycle cutset method consists of finding a minimal set of vertices in the constraint hypergraph such that the removal of these vertices leaves an acyclic hypergraph. Tree clustering consists of joining constraints in the problem in such a way that the resulting underlying constraint hypergraph is acyclic. These methods have been developed and implemented in various ways.

These techniques produce very effective algorithms. The only drawback is that they all require a heuristic step which can, in the worst case, lead to poor results. However, these are usually helpful techniques for hypergraphs which cannot be decomposed into small hinges.

Overall, the best results are obtained by first obtaining a hinge tree decomposition, and then solving each of the hinges using one of these heuristic approaches (Gyssens *et al.* 1994).

6.4.2 Relational structure

Recall again from Section 6.2.1 that abstracting away the scope from a constraint leaves a relation, and that we call a set of relations a constraint language. Another approach to determining the complexity of a class of CSPs is to analyse different constraint languages. In this way, we can utilize the relational structure of a CSP to bound its complexity.

For a given constraint language L, we define the constraint class $CSP(L)$ to be the class of CSPs whose constraints only use relations from L. We can then look for algorithms which efficiently solve all of the CSPs in $CSP(L)$.

If we find that our problem lies in some known $CSP(L)$ then we will have an upper bound on its complexity. One example of such a class of relations is the affine class, where each relation is expressed as a linear equation. Certainly there are efficient algorithms for solving such CSPs—we can even solve them optimally.

So, for different languages L, we can ask about the complexity of solving CSPs in $CSP(L)$. For bi-valued constraints (constraints where the maximum domain size is two) all of the languages for which there exist polynomial solution algorithms have been characterized (Schaefer 1978).

Since certain kinds of relation occur frequently in CSPs we can be hopeful that many CSPs will have a relational structure whose underlying complexity is known.

For constraint languages there are two main results, but, before stating them, we need some definitions, which rely on the fact that a solution to a CSP, P can be viewed as a relation $Sol(P)$ over the variables of P.

Definition 6.13 *A relation s* includes *a relation t if there is some projection of s which, after reordering, equals the relation t.*

Definition 6.14 *We say that a constraint language L can* express *a particular relation r if there is a CSP over L whose solution includes r.*

The expressiveness *of a language L is the set of all relations that can be expressed over L.*

The importance of the definition comes from the fact that, if a CSP, P, is given such that every relation of P can be expressed over a language L then that CSP is equivalent to a CSP, P' in CSP(L), in the sense that the solution to P' includes the solution to P.

The first important result it that we can determine the expressiveness of a constraint language from simple algebraic properties of its constituent relations. In fact, given a language L we can determine effectively whether any r can be expressed over L, and when it can, we can exhibit a CSP that bears witness to the fact (Cohen *et al.* 1996; Jeavons *et al.* 1997).

The second result is that the complexity of the constraint language is fully determined by simple algebraic properties of the relations involved, and can in fact be determined by solving a particular *Indicator Problem* which is a CSP defined over the language of interest (Jeavons *et al.* 1996).

6.4.3 Summary

We have shown that we can use the hypergraph structure and Graph Theory, or the relational structure and algebraic techniques to find efficient algorithms for CSPs.

In fact, given the hypergraph structure of a CSP we have an optimal decomposition and very effective heuristics. Given a constraint language we can often identify its complexity class exactly.

However, it is the interaction between the hypergraph structure and the relational structure that needs to be understood before a complete theory of CSP complexity can be derived. This is an important area for future research.

6.5 Prejudices about CSPs

In this last section we consider some aspects of constraints which go against natural intuition. Constraint satisfaction is a mature discipline and many good algorithms, techniques, and much incisive theory exists. However, there are still many prejudices around which are not supported by facts.

6.5.1 'Binary representations are better'

Recall the dual problem F_3 of Section 6.3.2.4. It is a binary CSP. All dual problems are binary CSPs. However, the cost is that the domain structure of the new variables encodes the structure of the original higher order constraints. Of course, exactly the same is true of the redundant variables method.

So, is it true to state that binary representations are adequate for every problem? Certainly every CSP can be expressed as a binary CSP but the constraint language involved is often very complicated. In fact, the encoding process often obfuscates the original problem structure. It is certainly not clear that the binary CSP is a better representation. It may not even be as good in general.

To illustrate this, consider the pigeon-hole problem. In this problem, we have n variables and the requirement that they all have different values. If the domain size is at least n then this is trivial to solve. If the domain size is smaller than n then there are no solutions.

This problem can be represented as a binary CSP in an obvious way. We simply constrain all pairs of variables with the constraint that they must take different values.

If this constraint appears as part of a larger CSP then this breaking up of the problem into a binary representation has obscured the simple nature of the original problem. There are constraint solvers which first search CSPs for configurations of this kind and solve them independently. This is quite efficient. However, if the constraint solver does not do this then the symmetry of the binary representation often causes many problems. Standard backtrack needs to search many different partial assignments before it can deduce the insolubility of any pigeon-hole problem.

It should be clear from this discussion that it is best to leave essentially higher order information in its original state. Breaking it into binary constraints is always possible but not often useful. The prejudice that binary constraints are better comes from the confidence in the strong techniques available in Graph Theory. It is a prejudice to be wary of. In the FAP we have to be clear that information is truly binary before we try to use binary representations.

6.5.2 'Weak constraints make solutions easier'

The second prejudice that we will consider concerns the *weakness* of a constraint. A constraint may be considered undemanding if it imposes very few restrictions on the values that may be taken by the constrained variables. The most undemanding a constraint can be is given by the following definition.

Definition 6.15 *A* quadrangle *is any constraint whose constraint relation is a direct product.*

This means that, aside from restricting the values of individual variables, the constraint imposes no joint condition on the variables. Valid assignments for this constraint may be chosen one variable at a time without any need ever to backtrack. (One way to view unconstrained variables is that they are in fact constrained by the quadrangle which is the direct product of their domains.)

Certainly CSPs whose language consists only of quadrangles are very easy to solve. However, we only need to extend quadrangles very slightly in order to obtain intractable constraint languages.

Definition 6.16 *A* subquadrangle (Rodosek 1997) *is a constraint whose constraint relation r has the property that any proper projection of r is a quadrangle.*

Subquadrangles are very weak in that any proper subset of the scope can be assigned values without recourse to backtracking. In other words we can assign values to all but one of the constrained variables in any way we like, and in any order, and we guarantee that there will be a value for the last variable in the scope which extends this assignment to a valid assignment of the constraint.

There are many examples of subquadrangles. For instance

1. All of the FAP constraints of Example 6.8 are subquadrangles.
2. All binary constraints are subquadrangles.
3. All of the scheduling constraints of Example 6.1 are subquadrangles.

These examples indicate that we can express scheduling problems and the FAP using only subquadrangles. Perhaps our intuition that weak constraints lead to CSPs which are easy to solve is wrong.

The following theorem puts pay to the idea that the constraint language consisting of all subquadrangles is tractable. We need some definitions before we can state the theorem.

Definition 6.17 *The* box *of a constraint c is the smallest quadrangle over the same scope as the constraint which allows all labellings allowed by c.*

Definition 6.18 *The* intersection *of two constraints c_1 and c_2 over the same scope s is a constraint with scope s whose constraint relation is the intersection of the constraint relations of c_1 and c_2.*

Theorem 6.19 *Any constraint c is the intersection of two subquadrangles with the same box as c.*

What this theorem means is that any CSP language can be expressed using subquadrangles. What is more, this can be done without altering the set of allowed values for any variable taken independently.

This very strong result puts pay to the notion that weak constraints are easy.

6.6 Conclusion

CSPs are a very widely applicable paradigm for representing and solving problems. Furthermore, there are many different representations of a particular problem as a CSP. In this chapter we have seen

(1) Pre-processing techniques to reduce redundancy in a CSP representation.
(2) Reformulation techniques which completely change the structure of a CSP.
(3) Certain constraint languages expressive enough to capture all relations.

It is apparent that modelling is a crucial part of using the CSP paradigm effectively. Modelling must capture the structure of the actual problem in the CSP. In particular the scope for the constraints must be appropriate. While any CSP may be transformed into a binary CSP, this is often at the cost of obfuscating original problem structure.

Once we have properly captured relevant problem structure in the structure of a CSP it then follows that good solution techniques must utilize the structure of the CSP. Of course, this applies both to the hypergraph structure and the relational structure of the CSP.

7

CELL AND FREQUENCY PLANNING

Stuart Allen, Stephen Hurley, and Roger Whitaker

7.1 Introduction

The engineering of radio networks for mobile communications consists of several tasks—the estimation of a traffic demand model, base station site selection, antenna configuration to achieve area coverage while minimizing the risk of interference and ultimately channel assignment (in FTDMA networks).

The network engineering task is obviously a precursor to the well known and much studied channel assignment problem (CAP). In this problem channels must be assigned to particular transmitters subject to certain objectives. The CAP is often represented using a graph colouring model and hence usually expresses minimum required channel separations between pairs of transmitters in order to reduce interference on receivers in the radio network to acceptable levels.

There are clearly many ways in which the requirements of an operational radio network can be interpreted into a set of channel separation constraints to be solved in the CAP. However, additionally there are many ways in which antennae in the radio network can be sited and configured. The broadcast power of an antenna produces a region of coverage, that is a region in which the radiated signal power is received by listening equipment with sufficient power to be useful. By convention in cellular networks receiving equipment is said to select the transmitter with the strongest received power. Hence, each transmitter forms regions which can be considered its sphere of influence. This set of regions is referred to as the *cell plan*, and the process of designing a cellular network as *cell planning* (see Fig. 7.1).

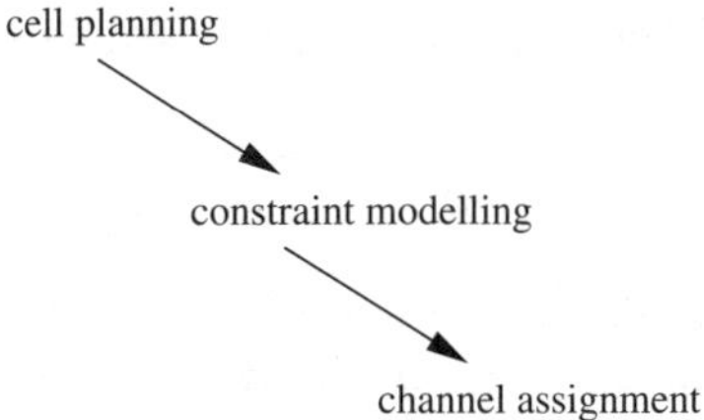

FIG. 7.1.

The cell planning problem (also known as the radio coverage problem, the antenna positioning problem or the network design problem) can be informally described as follows—given a list of candidate sites for transceiver antennae, a list of antenna types, and a discretized geographical working area characterized by a set of points with information relating to traffic estimation, service thresholds, and propagation, the aim is to select some sites among the candidate sites, and for each selected site determine the number and types of antennae, as well as associated parameters for each antenna. These decisions must satisfy some objectives which usually relate to area coverage and capacity.

7.2 Problem description

In this section we will briefly describe the planning problem; full details of the model used can be found in (Hurley 2000). A cellular radio network is composed of three objects:

(1) a discretized geographical *working area*, where signals and traffic are measured;
(2) *receiver equipment*, for example mobile telephones, which define the service requirements within the working area;
(3) *antennae*, which can be located on some predefined sites within the geographical working area.

7.2.1 Working area

A number of points are defined on the working area:

1. A set, R, of *reception test points* (RTP) at which propagation information is recorded.
2. A set, S, of *service test points* (STP), where the radio signal must be higher than a specified threshold for example $-90\,\mathrm{dBm}$ (the exact threshold is based on the type of receiver equipment at the STP).
3. A set, T, of *traffic test points* (TTP), which gives the traffic demand, measured in Erlangs, at the point. This measures the peak capacity requirements at the point.
4. A set, Z, of candidate sites at which antennae could be placed. It is assumed in this work that at most three antennae could be placed at any particular candidate site (different values could be used as appropriate).

Typically, we have

$$T \subseteq S \subseteq R.$$

Usually R forms a rectangular grid. Figure 7.2 shows an example region of a typical data map. The candidate sites are shown as white dots, the RTP set as the black rectangular region, the lightest grey areas coincide with STP and the slightly darker grey areas correspond to TTP.

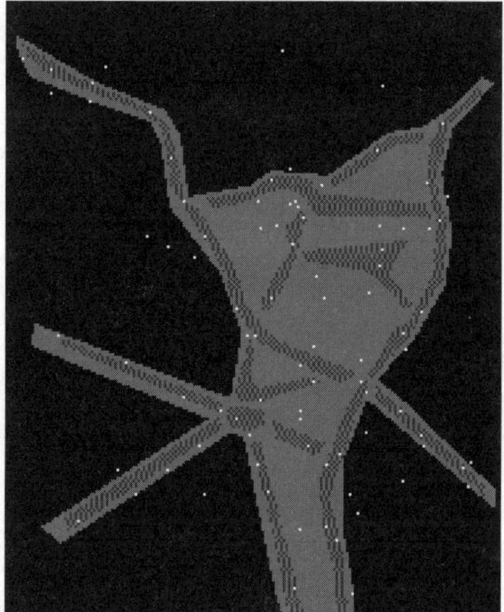

FIG. 7.2.

TABLE 7.1.

Type of Service	Threshold S_q (dBm)
8 W outdoor	−90
2 W outdoor	−83
2 W in car	−82
Indoor	−75
Deep indoor	−65

7.2.2 Receiver equipment

A network provides a service for all receiver equipment represented in the working area. A service threshold, S_q, defines the required level of service at each STP (which can vary throughout the working area if different services are provided). The exact value of the threshold at a STP is dependent on the equipment at the point. Examples are given in Table 7.1.

7.2.3 Antennae

Many types of antennae exist with differing characteristics such as radiation patterns and transmission gains and losses. In this chapter, we will assume that three types of antenna are available that could be deployed at a candidate site.

TABLE 7.2.

Antenna	Gain (dBm)	Loss (dBm)
Omni	11.15	7.00
Small	17.15	7.00
Large	15.65	7.00

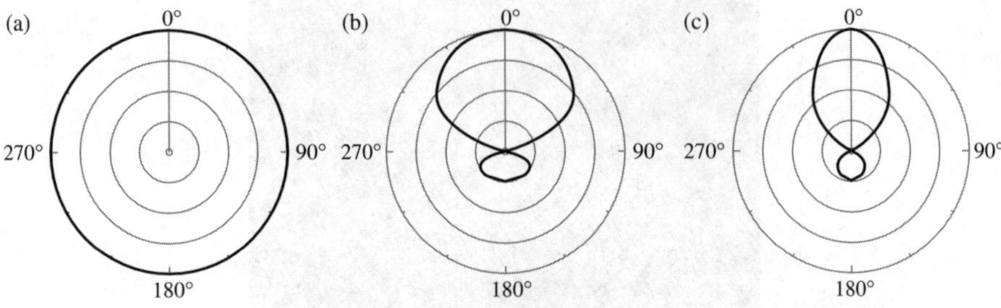

FIG. 7.3.

These are

(1) an *omnidirectional* antenna (omni);
(2) a *small directional* antenna (small); and
(3) a *large directional* antenna (large).

The associated gains and losses of these antennae are shown in Table 7.2. For illustration the horizontal radiation patterns for the omni, large and small antenna is shown in Fig. 7.3(a), (b), and (c) respectively.

To *configure* an antenna, values for power, azimuth (horizontal direction of antenna relative to North), and tilt (vertical angle of the antenna, relative to the horizontal) need to be allocated by the cell planner (whether manually or automatically by an optimization system). The ranges for each of these values,[1] used in this work, is as follows:

(1) Power: 26–55 dBm, in steps of 1 dBm;
(2) Azimuth: 0–359° (for directional antennae only), in steps of 1°;
(3) Tilt: 0 to −15° (for directional antennae only), in steps of 1°.

7.3 Network requirements

When modelling a cell plan two distinct issues need to be addressed. These are the *RF requirements* and the *teletraffic capacity requirements*. Most of the papers

[1] The values could vary from one network design to another.

on cell planning have paid considerable attention to coverage problems. That is, designing cellular networks which ensure that all STP are served by at least one antenna with adequate signal levels so that area coverage requirements are satisfied. Economic factors are then portrayed by minimizing the number of sites and antenna used.

In practice, cellular networks are *capacitated*, that is, the level of traffic that can be assigned to a particular antenna is limited. This generally forces a limit on the cell size and increases the number of required antenna sites. In FTDMA systems such as GSM each channel can support up to 8 calls, hence in a busy street there is likely to be a requirement for two or even more channels to cope with the anticipated traffic. The number of channels that an antenna can emit is limited by the technology involved, hence the size of high traffic cells may have to be reduced despite that the RF environment could provide coverage over a much wider area.

7.3.1 *RF requirements*

The RTP all have received path (propagation) loss data, in dBm (in the form of a *propagation loss matrix* (PLM)) and angle of incidence data (in the form of an AI matrix) which defines the vertical angles at which RTP appear to each site. In other models a path loss propagation function is computed at runtime (the models generally invoke a line-of-sight inverse power law to predict power loss with distance, for example Anderson and McGeehan (1994), or may be more complicated, for example Lattice–Boltzman models (Calégari *et al.* 1997*b*)).

The PLM and AI data enables a basic path loss calculation to be made for an antenna with a particular configuration, for example see Fig. 7.4 in which a link between a mobile user and an antenna is illustrated. In Fig. 7.4 the antenna is on a hillside radiating a signal down towards the user. However the free-space path is affected by terrain (and in this case some vegetation). The line of sight path loss which occurs in the PLM matrix referencing these points accounts for this. Further, the angle of incidence to the mobile differs from the tilt setting of the antenna and so vertical losses dependent on the antenna type are also incurred. A similar loss may be incurred on a horizontal setting if the mobile is not directly in line with the azimuth setting of the antenna (though this is not shown in Fig. 7.4).

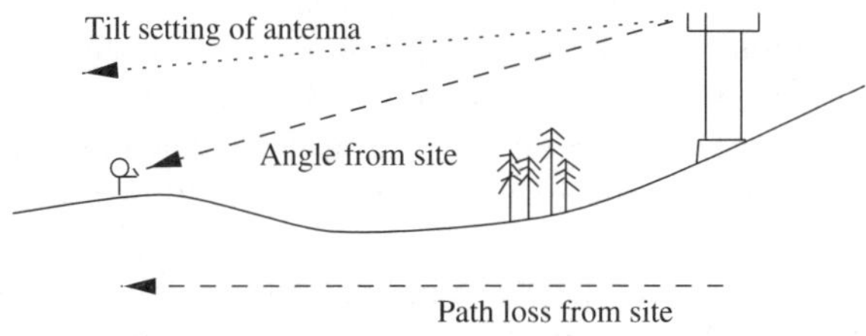

FIG. 7.4.

The calculations involved are now described in detail. The received power at an RTP (i.e. receiver), r, from an antenna, A, on a candidate site is calculated as follows

$$P_r = P^A + G^A - L_s^A - path\ loss_{rA} - L_h^A - L_v^A,$$

where

1. P_r is the received power (signal strength) at the RTP (in dBm).
2. P^A is the power setting of the selected antenna (in dBm).
3. G^A is the transmission gain of the selected antenna (in dB).
4. L_s^A is the transmission loss of the selected antenna (in dB).
5. $path\ loss_{rA}$ is a value reflecting propagation losses involving distance, fading over obstacles etc. between the receiver and the antenna (read from the PLM matrix) (in dB).
6. L_h^A is a loss value determined by the horizontal radiation pattern of the selected antenna (in dB).
7. L_v^A is a loss value determined by the vertical radiation pattern of the selected antenna (and uses the AI matrix) (in dB).

Definition 7.1 *A STP is* covered *if at least one antenna provides sufficient power that the service threshold is satisfied.*

Definition 7.2 *A* cell *is defined as the set of STP whose highest received power is from the same antenna (the* best server*) and above the required service threshold.*

The *cell plan* (*network design*) can thus be considered as a map of the region divided into areas (cells) which are associated with a particular best server. In Fig. 7.5(a) two antennae are shown to produce two cells, as denoted by the vertical and horizontal hatching on the mesh of reception points. In some cases these cells are disconnected (as is shown).

Definition 7.3 *A STP is said to be an* overlap STP *if more than one antenna provides a signal greater than the required service threshold at the point. The amount of overlap at a STP corresponds to the number of antennae providing an adequate signal at the point.*

It is desirable to have some overlap STP near the boundaries of cells to allow for handover requirements to be satisfied, however too much overlap at an STP or too many overlap STP increases the potential for interference at the points. In Fig. 7.5(b) we see that some STP are overlap STP (the mesh points with both horizontal and vertical hatching) that is both T_{a} and T_{b} supply sufficient signal strength to offer a service.

A region is *covered* if all the STP in that region are covered. Coverage is dependent on the selection of sites for placement of antennae. For each antenna placed, a power level, azimuth, and tilt must then also be specified. Once all

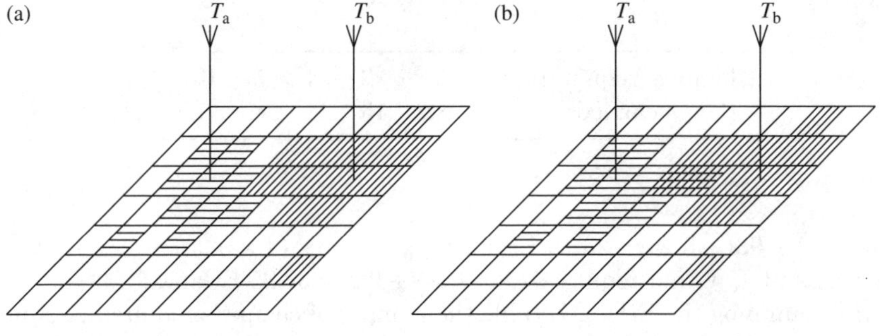

FIG. 7.5.

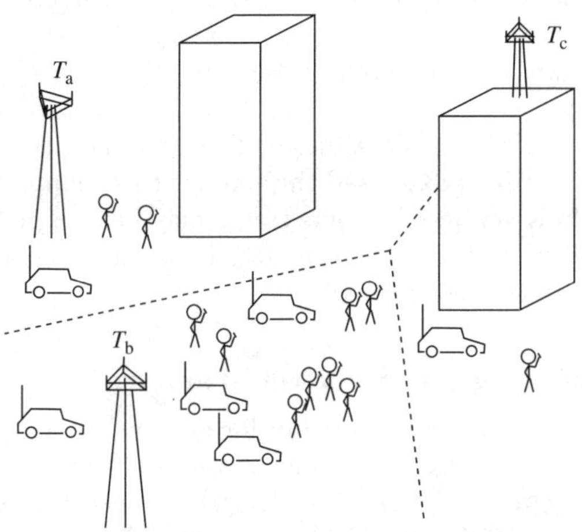

FIG. 7.6.

antennae are configured the propagation of signals can be calculated and the antenna providing the highest signal level becomes the best server at each RTP. If the sum of the traffic on the TTP associated with any particular best server exceeds a defined limit, then the *capacity* of that server is exceeded and the excess traffic would, in practice, be dropped as indicated in the next section.

7.3.2 Teletraffic capacity requirements

The number of channels available to an antenna is limited, hence also is the traffic that can be successfully serviced by the antenna.

Referring to Fig. 7.6, assume that a person with a mobile handset has an anticipated traffic demand of 0.2 Erlangs, a vehicle has a demand of 0.25 Erlangs, the leftmost building has a demand of 24.5 Erlangs, and the right most building has a demand of 37.8. The cell associated with T_a thus requires a total of

TABLE 7.3.

Channels required	1	2	3	4	5	6	7
Erlangs carried	2.9	8.2	15	22	28	35.5	43

$24.5 + 0.2 + 0.2 + 0.25 = 25.15$ Erlangs; T_b requires $(4 \times 0.25) + (8 \times 0.2) = 2.6$ Erlangs; and T_c would require $37.8 + 0.25 + 0.2 = 38.25$ Erlangs.

By examining the Erlang-B table the number of channels required to support the varied traffic demands can be found. An antenna can support traffic demands of up to 43.0 Erlangs. These traffic capacities refer to the number of channels required to support that demand *in the absence of interference*. These are shown in Table 7.3.[2]

Thus for the example illustrated by Fig. 7.6, T_a would require five channels, T_b would only need one channel, while T_c would need to be assigned seven channels. If the sum of the traffic values of TTP in a cell (its *traffic load*) exceeds the maximum given value (43 is used throughout this chapter), traffic cannot be guaranteed to be fully serviced or some traffic may be dropped, that is users will experience a 'busy' tone. The amount by which an antenna exceeds its maximum capacity limit is termed the *overload*.

7.4 A three-phase algorithm for cell planning

In this section, we describe an algorithm for generating cell plans that satisfies coverage and capacity requirements. The system implementing this algorithm is called DesNet (Design Network). The algorithm consists of three phases, each performed once in order. The three phases are as follows:

Initial design—aims to generate an initial, reasonably good cell plan which can be processed further. This is achieved by placing a few key antennae in such a way that a large degree of the region is covered (preferably with little overlap) but pays no attention to traffic capacity requirements.

Transition design—aims to process the cell plan from the initial design phase and improve its coverage and capacity performance reasonably quickly.

Final design—aims to ensure that the resulting cell plan is *feasible*. A feasible cell plan provides signal strength at each STP that satisfies its individual required service threshold, and also that the traffic load on each antenna is ≤ 43 Erlangs, that is 100 per cent coverage and 100 per cent capacity. The traffic load on an antenna is the sum of traffic required on TTP from which an antenna is its best server. Some protection against predicted interference is made though it is not a primary goal.

[2] It sometimes occurs that 58 Erlangs are permitted by adding an extra channel to be assigned.

7.4.1 Initial design phase

A number of initial design procedures have been tested, based on simple heuristics. The common underlying theme of these routines is that TTP should be covered where possible, the reasoning being that if TTP are covered then STP coverage is often attained. It is, therefore, desirable that antennae are placed on, or in close proximity to TTP. This has the additional effect of ensuring that when power levels are optimized in later phases, signal levels are highest on TTP—this helps ensure that the carrier to interference ratio at these points is high (which aids channel assignment). The following initial design procedures are available:

I1. This checks each candidate site (from a random ordering): if co-located with a TTP then activate it. The site initially has three sectors each with a large directional antenna. The first antenna is set with a random azimuth, subsequent antennae are separated by 120°. The pseudo-code is as follows:

```
for i := 1 to |Z| do
  if site[i] co-located with a TTP then
    if TTP is not covered (i.e. best signal is < service threshold) then
      activate a large directional antenna with random azimuth at site,
      activate a large directional antenna with 120° deflection,
      activate a large directional antenna with 240° deflection.
    end if
  end if
end for
```

After this initial placement the tilts and powers (all initially set at 0° and 50 dBm respectively) are dissipated—increased if the antenna carries less than 75 per cent traffic, or decreased if overloaded. Finally antennae carrying less than 8.2 Erlangs traffic are removed.

I2. This checks each candidate site (from a random ordering): if the selected site is on the same, or neighbouring to a mesh point representing a TTP and if the TTP is not already served by an existing antenna (i.e. the current best server provides a lower level of signal power than is required), activate an omnidirectional antenna at the site. The pseudo-code is as follows:

```
for i := 1 to |Z| do
  if site[i] is co-located with TTP then
    if TTP is not covered (i.e. best signal is < service threshold) then
      activate an omnidirectional antenna at 50 dBm on the site
    end if
  end if
end for
```

I3. This checks each TTP: if the currently considered TTP is not served by an existing antenna, activate the geographically nearest candidate site. The pseudo-code is as follows:

```
for i := 1 to |T| do
  if TTP[i] is not covered (i.e. best signal is < service threshold) then
    find the geographically closest candidate site co-located with a STP,
    activate an omnidirectional antenna at 50 dBm
  end if
end for
```

I4. This is similar to I3 but uses STP instead of TTP. In rural and quasi-urban networks there are often large spaces in the region which do not have TTP. Traffic based initialization methods such as I3 often fail to provide initial coverage at these points. The pseudo-code is as follows:

```
for i := 1 to |S| do
  if STP[i] is not covered (i.e. best signal is < service threshold) then
    find the geographically closest candidate site co-located with a STP,
    activate an omnidirectional antenna with power 50 dBm.
  end if
end for
```

I5. I3 often densely packs sites resulting in high levels of interference across the network, and the use of an exclusion zone spreads the sites throughout the region resulting in greater initial coverage of the region. I5 is similar to I3 but rejects any candidate site which is within a user supplied distance (the reuse distance) from any existing active antenna. Experiments have been conducted for reuse distances of 5, 8, 10, and 15 km. The pseudo-code is as follows:

```
for i := 1 to |S| do
  if STP[i] is not covered then
    find the geographically closest candidate site > reuse distance from any
    other already active site,
    if site is co-located with an STP then
      activate an omnidirectional antenna at 50 dBm
    end if
  end if
end for
```

I6. This randomly selects sites up to the theoretical minimum number of sites[3] necessary for a given network, given by

$$N_{\text{sites}}^{\text{min}} = \left\lceil \frac{1}{3 \times 43} \sum_{i=1}^{|T|} TTP[i] \right\rceil .$$

[3] The minimum number of antennae necessary, $N_{\text{antenna}}^{\text{min}} = \left\lceil \frac{1}{43} \sum_{i=1}^{|T|} TTP[i] \right\rceil$.

Fig. 7.8.

Fig. 7.9.

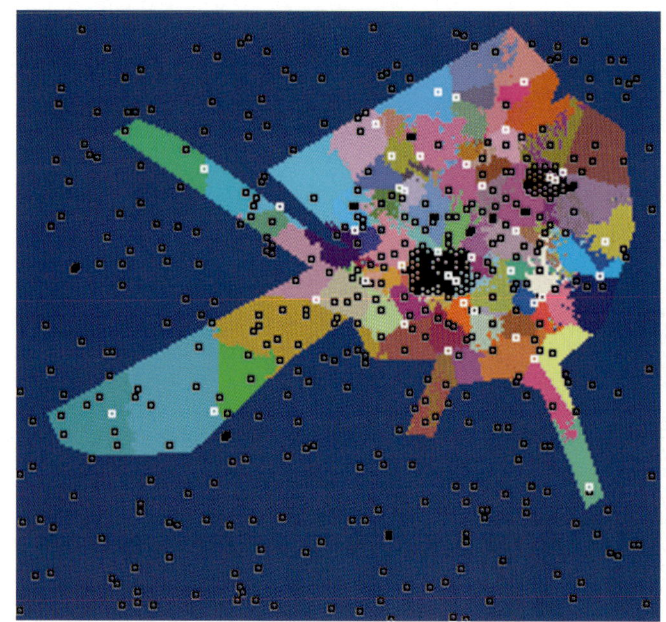

Fig. 7.10.

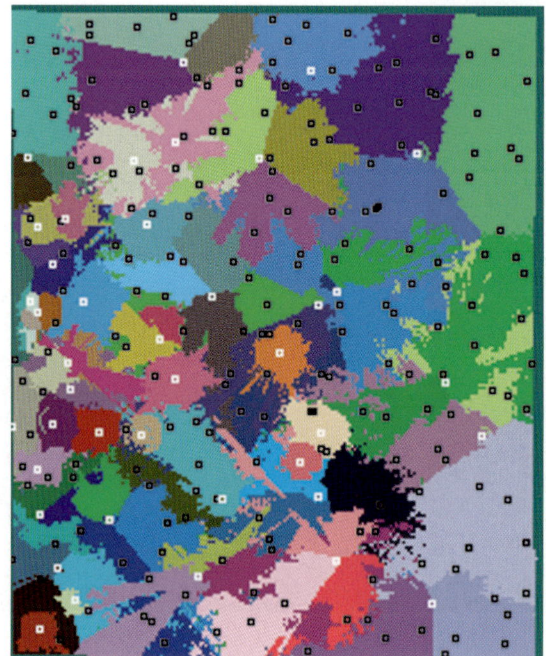

Fig. 7.11.

A simple randomized search then attempts to replace antenna on particular sites as follows. Each of the initially placed antenna is assumed to be omnidirectional but with a theoretical capacity of 129 Erlang (i.e. 3×43 Erlang). The search uses a simple cost function which sums the number of uncovered STP, the number of uncovered TTP, and the overload on each antenna (i.e. the sum of the overloads for traffic loads greater than 129).

The aim is to allow the search to produce a good initial network at minimal computational expense. Some probabilistic approaches tend to congregate sites on clusters of TTP resulting in cell plans with unnecessarily many active sites, and the 'spreading' approach (i.e. that of I5 which utilizes a reuse distance) shows that the reuse distance is difficult to set appropriately and would conceivably alter for every cell plan be designed. The pseudo-code is as follows:

```
for i := 1 to N_sites^min do
    activate a randomly chosen (inactive) candidate site with a random
    omnidirectional antenna
end for
for i := 1 to num_iterations do
    calculate cost of current cell plan, cost_old,
    deactivate a randomly selected active antenna, T_a,
    activate a randomly selected inactive antenna, T_b,
    calculate cost of new cell plan, cost_new
    if cost_new ≤ cost_old then
        retain network settings
    else
        reset T_a as active and deactivate T_b
    end if
end for
```

More sophisticated initial design procedures could be employed for example a genetic algorithm of the sort described by Calégari *et al.* (1997*a*,*b*) or the combinatorial method described by Molina *et al.* (1999).

7.4.2 Transition design phase

The transition design phase takes a cell plan from the initial design phase and applies four stages, once, in sequence. Each of the four stages is described as follows:

T1. (Sectorization). One of the main processes is the sectorization of sites. In T1 the potential for interference is reduced by replacing some omnidirectional antenna at a site with one, two or three directional antennae. Note that directional antennae are able to radiate emissions with greater gain in specific directions, and sectored sites can carry three times more traffic while maintaining coverage.

If a site carries more than 86 Erlangs it is sectored into a three-sector cell (e.g. an omnidirectional antenna is replaced by three large directional antennae and given random azimuth settings and 120° spacing). If the site carries more than 43 Erlangs but less than 86 Erlangs, it is replaced by two large directional antenna, and so on. This is shown in the following pseudo-code:

```
for all active sites do
  compute site-traffic-load i.e. traffic load on all antennae on the site
  if site has two antennae then
    if site-traffic-load > 86 Erlang then
      activate an additional randomly configured large directional antenna
    end if
  end if
  if site has one antenna then
    if site-traffic-load > 43 Erlang and site-traffic-load ≤ 86 Erlang then
      activate a randomly configured large directional antenna
      if antenna type is omni then
        deactivate omnidirectional antenna,
        activate a randomly configured large directional antenna.
      end if
    end if
    if site-traffic-load > 86 Erlang then
      activate two randomly configured large directional antennae
      if antenna type is omni then
        deactivate omnidirectional antenna,
        activate a randomly configured large directional antennae.
      end if
    end if
  end if
end for
```

T2. (Dissipate Antenna Azimuths). After sectorization, antennae azimuths are considered. Altering antenna azimuths can be one of the most effective means of producing cell plans to test. The azimuths of the antennae are deliberately altered in a continuous sweep of the possible values (in 10° increments). If the antenna originally carried more than 43 Erlangs and after the change carries less the azimuth is accepted. If the antenna originally carried less than 43 Erlangs but after carries more traffic but still less than 43, the change is also accepted.

```
for all active antennae do
  if antenna-type is small or antenna-type is large then
    compute antenna-traffic-load i.e. traffic load on antenna
    copy current-azimuth to local-azimuth
    for a := 1 to 36 do
      set antenna azimuth to (a * 10)
      compute new-traffic-load on antenna
      if antenna-traffic-load > 43 then
```

```
        if new-traffic-load ≤ 43 then
          local-azimuth := (a ∗ 10) {an improvement}
          antenna-traffic-load := new-traffic-load
        end if
      else
        if new-traffic-load ≥ antenna-traffic-load and
        new-traffic-load ≤ 43 then
          local-azimuth := (a ∗ 10) {an improvement}
          antenna-traffic-load := new-traffic-load
        end if
      end if
    end for
    set azimuth := local-azimuth {set to best found}
  end if
end for
```

T3. (Dissipate Down Tilt). If an antenna carries more than 43 Erlangs its down tilt is increased in steps of 1° until either the maximum tilt is reached (−15°) or the carried traffic drops below 43 Erlangs. Typically increasing the antenna tilt decreases the cell size and hence decreases the traffic carried.

```
for all active antennae do
  compute antenna-traffic-load i.e. traffic load on antenna
  while antenna-traffic-load > 43 do
    if tilt ≥ −14 then
      tilt = tilt−1 {do not tilt below −15}
      compute antenna-traffic-load
    else
      force exit from while loop
    end if
  end while
end for
```

T4. (Dissipate Powers). This increases the power of antennae that carry a low amount of traffic (arbitrarily set at 75 per cent of 43). Also for any antenna carrying more than 43 Erlangs its power is reduced in steps of 1 dBm until either the minimum power level is reached or the carried traffic drops below 43 Erlangs.

```
for all active antennae do
  compute antenna-traffic-load i.e. traffic load on antenna
  {dissipate power upwards}
  if antenna-traffic-load ≤ (0.75 ∗ 43) then
    while antenna-traffic-load ≤ (0.75 ∗ 43) do
      if power ≤ 54 dBm then
        increment power {do not use powers > 55 dBm}
        compute antenna-traffic-load
```

```
      if antenna-traffic-load > 43 then
        decrement power
        force exit from while loop
      end if
    else
      force exit from while loop
    end if
  end while
end if
{dissipate power downwards}
if antenna-traffic-load > 43 then
  while antenna-traffic-load > 43 do
    if power ≥ 27 dBm then
      decrement power {do not use powers < 26 dBm}
      compute antenna-traffic-load
    else
      force exit from while loop
    end if
  end while
end if
end for
```

7.4.3 Final design phase

The emphasis of this phase is to ensure that a feasible cell plan is generated, that is one in which all coverage and traffic capacity requirements are satisfied. In particular, to ensure that all STP are covered, that is each STP receives a signal from a base station which meets the required service threshold, and to ensure that all capacity requirements are met, that is no antenna may have a cell such that the best server coverage carries more than 43 Erlangs of traffic.

A simple hillclimbing scheme incorporating targetted operators is employed. Given a collection of heuristic operators, a particular antenna or antenna site is selected and a randomly chosen heuristic operator is applied as a form of local search.

7.4.3.1 Network evaluation The final design phase requires that a candidate cell plan has an associated cost. This is a heuristic estimate of its *goodness* that enables cell plans to be compared and an improved cell plan accepted over another design.

The cost function in the final design phase considers the number of STP which are not served to their required threshold, and the sum of the antenna traffic overloads. If an antenna carries 53 Erlangs of traffic for example, but is limited to carrying 43 Erlangs, then a contribution of 10 is added to the final cost. The cost, C, which reflects the coverage and capacity requirements, is thus defined

$$C = \text{number STP not covered} + \sum_{\text{active antenna}} (\text{traffic dropped by antenna}).$$

In general, operator changes which improve the network, $C_{\text{new}} < C_{\text{old}}$, are always accepted. Changes which return the same cost as before are termed *sideways* moves. In general hillclimbing approaches it is often advantageous to accept these, however whether to do so is dependent on the operator used.

As a consequence of using the cost function two lists of antennae are returned, based on an evaluation of their traffic status. Any antenna carrying more than 43 Erlangs is *overloaded*, but also any antenna carrying less than 21.5 Erlangs is termed *underloaded.* The purpose of these lists is made clearer in the next section.

7.4.3.2 Targetting reconfiguration operators DesNet implements a set of lists which are used to steer the final design phase into making changes to the configuration of antennae (or sites) which are deemed to be problematic. Clearly a sectored site with each antenna holding, say, 40 Erlangs and suffering little interference is a configuration which it would be undesirable to alter. If an antenna is overloaded its configuration may be beneficially altered, also if an antenna carries a very low level of traffic it could be wasteful.

The cost computation generates *underload* and *overload* vectors which are used by the search procedure to target areas which require improvement. The overload vector stores any antenna which carries more than the maximum permitted load of traffic. The underload and overload lists are then merged into a list called *choices.* If the choices list is empty then all active antenna are universally included. An antenna is chosen at random (uniform distribution) from the choices list and this antenna is the target for reconfiguration using one of the operators.

The overall final design procedure is given in Fig. 7.7. The termination criterion used for the procedure is to terminate if the overall improvement in cost did not exceed 2 per cent for every 100 steps (a step is an application of any operator). These values were determined experimentally and are a compromise between solution quality and execution time.

7.4.3.3 Final design operators The following heuristic operators are used in the final design phase; other operators could be used as appropriate depending on the requirements of the cell plan.

F1. DecrementPower—for the chosen antenna, if the power is higher than its permitted minimum setting, decrease the power (from its current value) by 1 dBm. If the resulting network cost is less or equal then accept it, otherwise return the power to its original value. A variation of this operator simultaneously decrements the powers of all overloaded antenna before assessment.

```
Compute cost of current network
while overload vector is non-empty and % coverage < 100 do
   combine underload and overload vectors into vector choices,
   randomly select an operator,
   select a random antenna from choices,
   apply operator to selected antenna,
   if operator improves network cost then
      recompute choices from new underload and overload vectors
   end if
end while
```

FIG. 7.7.

F2. IncrementPower—for the chosen antenna, if the power is lower than its permitted maximum setting, increase the power (from its current value) by 1 dBm. If the resulting network cost is less or equal then accept it, otherwise return the power to its original value. A similar variation to F1 exists.
F3. DissipatePowerUp—for the chosen antenna, while the traffic carried is less than 43 Erlangs, increase the power (from its current value) in steps of one dBm until either the upper power limit is reached or an unacceptable traffic level is carried (in which case step the power down one dBm). If the resulting network cost is less or equal then accept it, otherwise return the power to its original value.
F4. DissipatePowerDown—for the chosen antenna, while the traffic carried is greater than 43 Erlangs, reduce the power (from its current value) in steps of 1 dBm until either an acceptable traffic level is carried, or the lower power limit is reached. If the resulting network cost is less or equal then accept it, otherwise return the power to its original value.
F5. DissipateTiltDown—for the chosen antenna, while the traffic carried is greater than 43 Erlangs, increase the tilt (from its current value) in steps of $1°$ until either an acceptable traffic level is carried, or the maximum tilt limit is reached ($-15°$). If the resulting network cost is less or equal then accept it, otherwise return the tilt to its original value.
F6. AddNearbyAntenna—add a randomly configured antenna to a site (possibly unused) outside a given reuse distance from the chosen antenna. All antenna outside the distance but within twice the distance are considered and one is chosen at random (if no suitable antenna then the operator terminates). If the resulting network cost is improved the new antenna remains active otherwise it is removed, that is a sideways move on this operator is not permitted.
F7. SwitchAntennaType—if the antenna is omnidirectional then replace by a randomly selected directional antenna type. If the antenna is directional then

change to the alternative directional antenna. If the resulting cost is the same or improved then accept the change, otherwise return to its original setting.
F8. SimultaneousPowerUpTiltDown—while the power level on the selected antenna is less than its maximum permitted value, and the tilt is higher than its maximum permitted value (-15°), offset the power and tilt by one accordingly. If the resulting network cost is the same or improved accept the changes, otherwise reject.
F9. SimultaneousPowerDownTiltUp—while the power level is greater than its minimum permitted value, and the tilt is lower than its minimum permitted value (0°), offset the power and tilt by one accordingly. If the resulting network cost is the same or improved accept the changes, otherwise reject.
F10. ChangeAzimuth—if the selected antenna is not omnidirectional, test the full range of horizontal deflection in 15° steps, for example $0, 15, 30, \ldots, 330, 345$. If a new setting results in an equal or improved network accept it, otherwise return the azimuth to its original setting.
F11. AddRandomSiteForUncoveredTTP—find an uncovered TTP (that is one for which its best server signal is below the required service threshold), and activate a randomly configured antenna at its geographically nearest site holding less than three antennae. Only accept this change if the resulting network cost is an improvement.
F12. AddDeliberateStationForUncoveredTTP—scan uncovered TTP (that is one for which its best server signal is below the required service threshold), recording traffic load at the nearest site with less than three antennae and add a randomly configured antenna to the site. Only accept this change if the resulting network cost is an improvement.
F13. RandomlyReconfigure—randomly reconfigure the selected antenna (within applicable limits) for power, azimuth, and tilt. If the new configuration results in an equal or improved cost network then accept it, otherwise return the antenna to its original configuration.
F14. DeliberatelyReconfigure—reconfigure the antenna on the selected site (within applicable limits) for power, azimuth, and tilt (for azimuth attempt 24 random settings). If the new configuration results in an equal or improved cost network accept, otherwise return the antenna to its original configuration.
F15. RemoveIfCarryingLessThan—scan through all active antenna in the network. If an antenna carries less traffic than a user specified number of Erlangs (typically 2.9 Erlangs for 1 required channel, or 8.2 Erlangs for 2 required channels—as determined by the Erlang B table) remove the antenna. If the resulting network is an improvement accept it, otherwise reset the antennae to their original settings.
F16. DissipatePowerUpOnUncoveredSTPBestServer—randomly select an uncovered STP and determine the best server antenna, then while the traffic carried is less than 43 Erlangs, increase the power (from its current value) in steps of one dBm until either the upper power limit is reached or an unacceptable traffic level is carried (in which case step the power back down one dBm).

TABLE 7.4.

	Road	Rural	Town	Urban
$\|S\|$	29954	74295	17393	48512
$\|T\|$	4967	24999	6656	19703
$\|Z\|$	250	320	568	244
Size of region (km)	40×170	63×54	50×46	10×12
Mesh increment (m)	200	195	200	49
Service threshold (dBm)	−90	−90	−90	−90
Total required traffic capacity (Erlangs)	3210.9	2787.9	2988.1	2652.1
Minimum number of sites, $N_{\text{sites}}^{\text{min}}$	25	22	24	21
Minimum number of antennae, $N_{\text{antenna}}^{\text{min}}$	75	65	70	62

If the resulting network cost is less or equal then accept it, otherwise return the power to its original value.

7.5 Data sets and results

Four real world data sets have been used for study. All data sets were provided by CNET.[4] The four networks represent four geographic and demographic problems. The first, *Road*, characterizes a road network, *Town* considers the service requirements of a medium sized town, *Rural* represents a region containing a mixture of sparsely populated areas, an urban area and some roads, and finally *Urban* represents a region containing a dense urban area. The characteristics of each type of data set is given in Table 7.4.

7.5.1 Results of initial and transition phases

Table 7.5 records the results of the different initial design (ID) phases, each followed by the transition design (TD) phase[5] that is pre-final design, for the Road, Rural, Town, and Urban networks. The 'coverage' column gives the percentage STP coverage that is area coverage of the cell plan; the 'capacity' column gives the capacity of the designed network in Erlangs as a percentage of the total traffic capacity required; 'sites', 'omni', 'large', 'small', give the number of candidate sites, omnidirectional antenna, large directional antenna and small directional antenna used in the cell plan. The 'time' column records the execution time in minutes[6].

[4] France Telecom Research and Development Centre, Belfort, France.

[5] T1 is not used after I1 i.e. only T2, T3, and T4.

[6] All timings were carried out on a 550 MHz PC computer with 512 MB RAM running C++ program code.

TABLE 7.5.

Method	Coverage (%)	Capacity (%)	Sites	Omni	Large	Small	Time
Road network							
I1 + TD	91.67	47.63	16	0	29	13	36
I2 + TD	90.77	51.55	17	4	18	18	27
I3 + TD	98.74	72.45	27	6	41	15	47
I4 + TD	98.45	71.68	33	15	34	12	52
I5 + TD	98.57	72.31	27	7	35	20	57
I6 + TD	99.56	83.90	24	1	46	21	119
Rural network							
I1 + TD	89.94	75.11	16	0	29	17	92
I2 + TD	89.42	73.23	18	5	22	15	78
I3 + TD	99.69	88.01	52	32	35	13	145
I4 + TD	95.27	87.15	71	38	42	21	151
I5 + TD	99.73	88.86	52	31	38	11	162
I6 + TD	95.57	75.99	22	3	34	13	208
Town network							
I1 + TD	70.43	22.14	6	0	7	10	5
I2 + TD	64.92	26.80	7	0	5	14	6
I3 + TD	91.00	39.37	12	1	17	12	9
I4 + TD	92.35	48.15	15	2	21	13	11
I5 + TD	93.16	40.11	12	2	16	12	10
I6 + TD	99.84	84.43	24	2	46	15	62
Urban network							
I1 + TD	43.42	24.31	5	0	7	8	11
I2 + TD	71.74	33.95	7	0	11	10	18
I3 + TD	87.48	42.88	15	7	17	7	24
I4 + TD	89.43	44.44	16	7	13	13	30
I5 + TD	64.14	66.11	8	5	1	11	9
I6 + TD	99.98	99.98	49	27	38	11	123

It is observed from the results that initial design method I6 generally produces the best area coverage and capacity cell plans (except for the *Rural* network) but at a higher computational cost.

7.5.2 Results of final design phase

Table 7.6 records the results of the final design phase for the Road, Rural, Town, and Urban networks generated by the different initial design phases followed by the transition design phase. The strict requirement that all STP are covered is

TABLE 7.6.

ID	Sites	Antenna	Omni	Large	Small	Lost traffic (%)
Road network						
1	59	108	12	60	36	7.08
2	60	95	14	48	33	6.22
3	52	106	13	58	35	5.86
4	58	102	15	56	31	8.19
5	56	106	7	62	37	10.32
6	48	106	6	64	36	8.04
Rural network						
1	58	104	11	55	38	10.45
2	68	105	17	48	40	11.31
3	65	100	37	41	22	14.31
4	101	147	41	75	31	18.11
5	64	97	36	43	18	7.82
6	57	100	13	60	27	13.95
Town network						
1	54	89	8	49	32	21.32
2	50	89	10	51	28	15.74
3	50	93	10	46	37	21.97
4	62	101	18	51	32	17.70
5	54	100	8	47	45	20.71
6	39	85	3	49	33	12.03
Urban network						
1	51	89	9	48	32	9.65
2	52	88	15	45	28	12.56
3	50	87	16	45	26	11.18
4	57	96	15	48	33	17.70
5	50	85	14	43	28	17.92
6	41	79	14	40	25	10.76

satisfied for all networks, except for I1 on the Rural network (after an excessive amount of time in which no improvement of the search was completed some eight STP remained uncovered, all TTP were covered however). In each case all TTP are covered and the capacity of any antenna is never exceeded.

For each of the networks in Table 7.6 the column entries are for the initial design method, the number of sites used in the final cell plan followed by the total number of antennae placed. The number of the three different kinds of

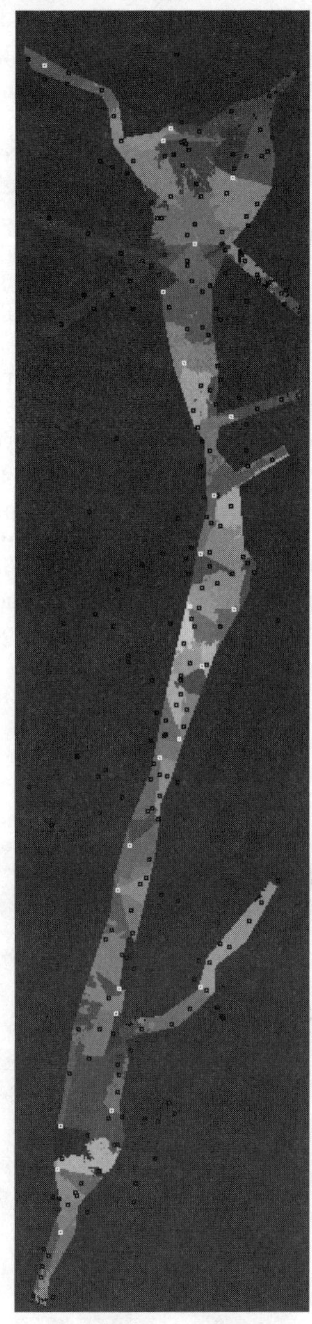

FIG. 7.8. See Plate 1.

antenna deployed in each cell plan are given in the 'omni', 'small' and 'large' columns. The 'lost traffic' column indicates the amount of traffic (in Erlangs) which is *expected* to be lost due to interference following a channel assignment.

Generating channel separation constraints from a given cell plan is a time consuming process and not unique, therefore to assess the ability with which a frequency plan could ensure that the capacity of a cell plan is actually available in an operational network, some intermediate measure of difficulty is useful. The 'lost traffic' column indicates such a measure—the simple interference measure used counts the number of received powers above a given offset of the highest received power. The measure used states that any RTP receiving seven or more signal levels which are greater than 9 dB down on the highest received power there will be too much interference at the point to provide a service whatever the final channel assignment (frequency plan).

Overall better results occur when using initial design method ID6 (markedly so for the Town and Urban networks, though less apparent for the Road and Rural networks). The method produces cell plans requiring the lowest number of candidate sites and for the Town and Urban networks the lowest number of antennae. However, it only produces the least traffic loss in the Town network.

Figures 7.8–7.11 (also Plates 1–4) illustrate the best cell plans produced by the three phase algorithm for the Road, Rural, Town, and Urban networks respectively.

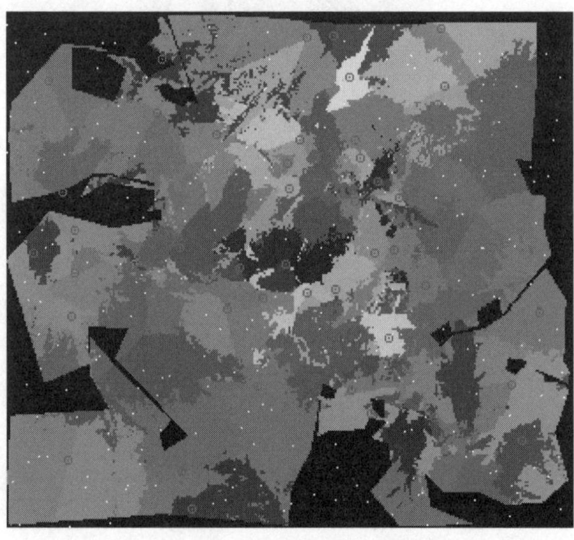

FIG. 7.9. See Plate 2.

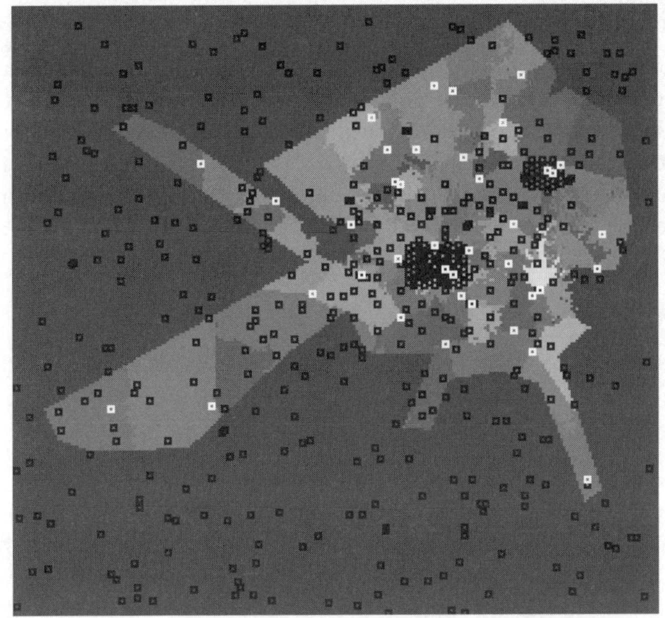

FIG. 7.10. See Plate 3.

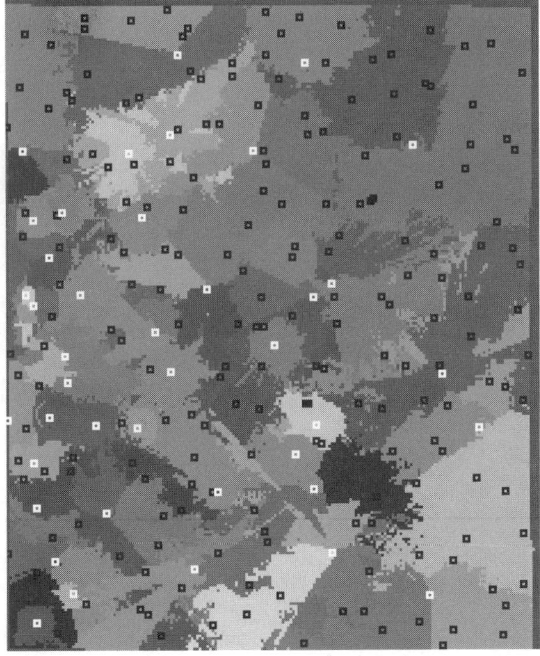

FIG. 7.11. See Plate 4.

7.6 Concluding remarks

A model for antenna site selection and antenna configuration has been presented which addresses area coverage and peak capacity requirements. A three phase algorithm is developed to generate cell plans with the required characteristics. Results show that the algorithmic process is successful and is able to generate effective plans. In particular it can be concluded that the cell planning problem is tractable.

8

ECONOMIC ASPECTS OF RADIO CHANNEL ASSIGNMENT

Robert Leese

8.1 Introduction

The previous chapters in this volume have mainly addressed the radio engineering aspects of channel assignment. In particular, they have considered, in various settings, the relationship between the amount of spectrum allocated to a radio service and the interference that is experienced by users of the service. The aim in this final chapter is to demonstrate how economic modelling may be integrated into such frameworks, with the result that spectrum assignments may be evaluated not just in terms of engineering measures such as interference levels or span, but also in terms of economic measures such as operator and consumer surpluses. One then has a means of estimating the economic value of radio spectrum in light of the combined engineering and economic environments in which it is deployed.

The model that is developed below has three key components: a regulator (e.g., the Radiocommunications Agency in the UK or the FCC in the US); a set of operators, who provide a radio service by using spectrum on terms that are determined by the regulator; and consumers, who purchase the service from the operators. Consumers have an elastic demand for the radio service, in response to the price at which it is offered. Operators are free to set the price of the service in an attempt to maximize their profits. The regulator is generally interested in maximizing some combination of consumer and operator surpluses, and encourages the market to reach such a point through various spectrum management tools. In this treatment, we shall consider a regulatory framework comprising the span of the spectrum allocation together with a simple pricing structure.

The remainder of the chapter is organized as follows. The next section expands on the broad model outlined above, explaining the addition engineering and economic assumptions that are made. Section 8.3 solves a set of small examples in order to highlight in a quantitative way the various economic incentives that are present. Section 8.4 briefly discusses more general situations that might be handled within a similar theoretical framework.

8.2 A combined engineering and economic model

8.2.1 *General framework*

We shall develop a model consisting of a set of markets, in which operators compete for customers. In order to simplify the analysis, it is assumed that there are n identical operators, who share the activity in every market equally. The markets are independent of each other and, again for ease of analysis, have the same variation of consumer demand X with price P, given by

$$X = \frac{A}{P^{\epsilon}}, \tag{8.1}$$

where A and ϵ are both constants, called the demand potential and the elasticity of demand, respectively. Note that P may vary from market to market, and the way in which operators reflect the local radio interference environment in their choices of P is one of the fundamental features of the model. Letting P be different in different markets means that supply of the radio service will always equal the demand. If there are s markets, denoted $M_1, M_2, \ldots, M_s$, then the supply X_i in market M_i is related to the price P_i by $X_i = A/P_i^{\epsilon}$. Note the assumption that A and ϵ are the same for all markets; this could be relaxed without any change to the present model, but it has the advantage of allowing a less cluttered presentation of the main conclusions. It is also assumed that $\epsilon > 1$, which is consistent with economic studies of radio services and means that demand falls off sufficiently fast as prices increase that operators naturally have incentives to provide a low-price service to a large customer base.

Since the n operators are identical, each one supplies at a level of $x_i = X_i/n$ and generates an income of $x_i P_i$. In keeping with the natural language of economics, we shall from now on call the operators 'firms' and speak of x_i as the 'output' of each firm in market M_i. The costs of the firms have two components: a constant marginal cost c for each unit of output and a fixed cost a for each radio channel to which the firm has access. If we think of x_i as the output *per unit time* then we can think of a as a fixed licence fee, payable to the regulator per unit time (e.g. annually) in respect of the channels that are subject to those licences. To combine these components into an overall profit function, we first need to have a clear description of the radio interference environment.

8.2.2 *Radio interference constraints*

The possible ways of formulating radio interference constraints within the channel assignment problem have been treated extensively elsewhere in this volume. In principle, any of them could be incorporated into an economic model, but even the simplest reveals key features of the interplay between engineering and economic aspects, and this is the path followed here. In particular, it is assumed that only co-channel interference is a potential hazard, meaning that the interference environment may be described by a simple, undirected graph. In contrast to the usual interpretation of nodes in the graph as transmitter sites, they should now be interpreted as markets, with the numbers of channels deployed in each to be determined later by economic considerations. Edges in the graph join markets that are prevented from using the same channel because of the unacceptable radio interference that would result.

To maintain flexibility in the number of channels used in each market, we think of a channel assignment as a (multi)-covering of the nodes in the interference graph by independent sets. (An independent set is a subset of the nodes, no pair of which are joined by an edge.) The pattern of independent sets in the interference graph captures the scope for spectrum reuse. In fact, it is necessary to consider only *maximal* independent sets, that is those independent sets to which no further nodes can be added. Suppose that the maximal independent sets are $\{I_k : k = 1, \ldots, t\}$; then a channel assignment can be built up by 'covering' the interference graph with a number of copies of each independent set, with each copy of each set set being associated with a distinct radio channel. If Y_k is the number of copies used of independent set I_k then the overall number of channels available in market M_i is $\sum_{\{k:M_i \in I_k\}} Y_k$.

8.2.3 *The optimization problem faced by firms*

We are now in a position to write down the profit function for each firm. Taking into account the marginal cost c, the income from market M_i is $(P_i - c)x_i$. The additional cost of spectrum access in respect of independent set I_k is ay_k, where $y_k = Y_k/n$ is the number of copies of I_k used by each firm. Hence, the profits of each firm are

$$\Pi = \sum_i (P_i - c)x_i - a\sum_k y_k \,. \tag{8.2}$$

In writing (8.2), the unit of output is implicitly assumed to be that made possible by a single radio channel. There is then a capacity condition that the output in market M_i cannot be larger than the number of channels available there, which is explicitly written

$$x_i - \sum_{\{k:M_i \in I_k\}} y_k \leq 0 \,. \tag{8.3}$$

We shall also consider the effects of allocating a limited overall amount of spectrum to the service. If there are r channels available per firm then

$$\sum_k y_k - r \leq 0 \,. \tag{8.4}$$

The optimization problem faced by the firms is, therefore, to choose the $x_i \geq 0$ and $y_k \geq 0$ to maximize (8.2), subject to the constraints (8.3) and (8.4). It may be treated as a Kuhn–Tucker problem, in which we introduce dual variables $\lambda_i, \tau_k, \mu \geq 0$; λ_i is associated with the constraint (8.3), τ_k is associated with the constraint $y_k \geq 0$ and μ is associated with the constraint (8.4). If, in the solution to the profit-maximization problem, any of these constraints do not bind, that is are satisfied with strict inequality, then the corresponding dual variable takes value zero. Conversely, if any of the dual variables take nonzero values, then the corresponding constraint binds. These are the so-called complementary slackness conditions. There is no need to explicitly introduce a dual variable for the constraint $x_i \geq 0$, since the demand curve (8.1) has the feature of arbitrarily high prices at low demand, meaning that the firms will never choose a zero output in any market.

We have assumed implicitly that y_k and x_i may take fractional values, which at first sight is inconsistent with the way in which spectrum is divided into discrete channels. However, radio channels are often divided between different users in the time domain, which creates a much finer division of radio capacity than a straightforward assignment of discrete channels in the frequency domain. We do not consider these effects explicitly, but instead bear them in mind as motivation for our formulation in terms of a continuous optimization problem.

To complete the description of the solution, it remains only to write down the usual first-order conditions in this type of problem for the variables x_i and y_k, which are respectively

$$P_i - c + \left(\frac{\partial P_i}{\partial x_i}\right) x_i = \lambda_i, \tag{8.5}$$

and

$$\sum_{\{i:M_i \in I_k\}} \lambda_i - \mu + \tau_k = a\,. \tag{8.6}$$

These equations are central to all of what follows.

8.2.4 Economic equilibria

To perform calculations in the above framework requires a more explicit expression for the partial derivative $\partial P_i/\partial x_i$ in (8.5). We shall, therefore, assume that the firms reach a Cournot–Nash equilibrium, corresponding to an assumption that firms do not anticipate changes in the output of other firms when choosing their own output. In other words, $\partial P_i/\partial x_i$ may be replaced by dP_i/dX_i as given by (8.1):

$$\frac{dP_i}{dX_i} = -\frac{P_i}{\epsilon X_i} = -\frac{P_i^{\epsilon+1}}{\epsilon A}\,, \tag{8.7}$$

which in turn means that (8.5) may be rewritten

$$\frac{x_i P_i^{\epsilon}}{A} = \epsilon\left(1 - \frac{\lambda_i + c}{P_i}\right). \tag{8.8}$$

Since the firms contribute equally to the output in each market, the left hand side of (8.8) is equal to $1/n$ and so it may be rearranged to give

$$P_i = \frac{\lambda_i + c}{1 - 1/(\epsilon n)}\,, \tag{8.9}$$

which has the same form as the familiar expression for equilibrium pricing in a Cournot–Nash oligopoly.

The influences of spectrum reuse and limited spectrum allocations on the market prices are both captured by λ_i, in ways that will become clear in the examples of the next section. Once the prices have been determined then the consumer and operator surpluses follow. The consumer surplus Γ_i in market M_i is the area underneath the graph of X versus P to the right of the market price P_i, which from (8.1) is

$$\Gamma_i = \frac{A}{(\epsilon - 1)P_i^{\epsilon-1}}. \tag{8.10}$$

The total operator surplus taken across all n firms is simply $n\Pi$, where Π is given by (8.2). If we add back in the spectrum fees $na\sum_k y_k$, on the grounds that they are a surplus that accrues to the regulator, then the total welfare W, defined to be the sum of consumer, operator, and regulator surpluses, may be written

$$W = \sum_i \frac{A}{P_i^\epsilon}\left(\frac{\epsilon P_i}{\epsilon - 1} - c\right). \tag{8.11}$$

8.3 Examples

8.3.1 *General remarks*

The following examples demonstrate the application of the above model to a number of small interference graphs. They are designed to highlight some of the qualitative features that emerge when the engineering and economic aspects of channel assignment are brought together. Of particular interest are the effects that the regulator can have through the choices of spectrum fee a and spectrum allocation r.

In all of what follows, the monetary units and units of output are fixed by setting $A = 1$ for the demand potential and $c = 1$ for the marginal cost of output. The first of these choices means that the output x_i of each firm in market M_i is related to the market price by

$$x_i = \frac{1}{nP_i^\epsilon}. \tag{8.12}$$

It is also a convenient shorthand notation to write

$$\gamma_n = \left(1 - \frac{1}{\epsilon n}\right)^{-1}, \tag{8.13}$$

so that the market prices become

$$P_i = \gamma_n(1 + \lambda_i). \tag{8.14}$$

For each example, there will be identified an *unrestricted* regime, in which the limited spectrum allocation r does not come into play, with the consequence that the dual variable μ is equal to zero, and a *restricted* regime, in which $\sum_k y_k = r$

and μ may be nonzero. We shall derive results for general parameter values, and display them in the particular case of three firms and an elasticity of demand equal to 1.2.

8.3.2 *Complete interference graph*

A complete graph has edges between every pair of nodes. The special property of a complete interference graph is that there is no possibility for spectrum reuse. There are precisely as many maximal independent sets as markets, that is $s = t$ with $I_i = \{M_i\}$ for $i = 1, \ldots, s$. Each I_i must be used in the channel assignment, since otherwise there will be some x_i with value zero, and therefore $\tau_i = 0$ for all i.

UNRESTRICTED REGIME: In the unrestricted regime, $\mu = 0$ and so (8.6) gives directly that $\lambda_i = a$ for all i. Hence all the market prices are equal, as one might expect. Since $\lambda_i > 0$, there is equality in (8.3), that is $x_i = y_i$ for all i. This regime will be the one of interest provided that (8.4) is also satisfied, which implies that $r \geq s/(nP_i^\epsilon)$.

RESTRICTED REGIME: The limited spectrum allocation affects the choice of output when $r < s/(nP_i^\epsilon)$. Compared to the unrestricted regime, market prices will be higher, since now $\lambda_i = \mu + a$ for all i. Since the λ_i are all equal to each other, the P_i and x_i are also the same for all i. Since $\lambda_i > 0$, we again have $x_i = y_i$, that is the y_i are also independent of i. The requirement that $\sum_i y_i = r$ means that $x_i = y_i = r/s$; the market prices then follow from (8.12), from which λ_i and then finally μ may be calculated.

A useful diagrammatic representation of the two regimes uses $(1 + a)^\epsilon$ and $rn\gamma_n^\epsilon$ as coordinates. The boundary between the two regimes in this example occurs when $s = rnP_i^\epsilon = rn\gamma_n^\epsilon(1 + a)^\epsilon$, that is when

$$rn\gamma_n^\epsilon = \frac{s}{(1 + a)^\epsilon}\,. \tag{8.15}$$

Figure 8.1 shows for $s = 4$ the total welfare per market as a function of $(1+a)^\epsilon$ (on the front axis) and $rn\gamma_n^\epsilon$ (on the right-hand axis). The surface is shaded so that higher values are lighter; the boundary between the two regimes is clearly visible as a curve from the rearmost corner of the plot to the foremost corner. The restricted regime is in the foreground, where welfare falls off sharply as the amount of allocated spectrum is reduced. Welfare is highest at the rearmost corner, where spectrum prices are low and the amount of allocated spectrum is high. In general, the needs of other services mean that a regulator does not have the luxury of making large spectrum allocations and so, in practice, a balance would be desirable, most likely near the boundary of the two regimes.

8.3.3 *Three nodes and two edges*

The next two examples relate to the three-market interference graphs shown in Fig. 8.2. In the first, there are two edges and the maximal independent sets are

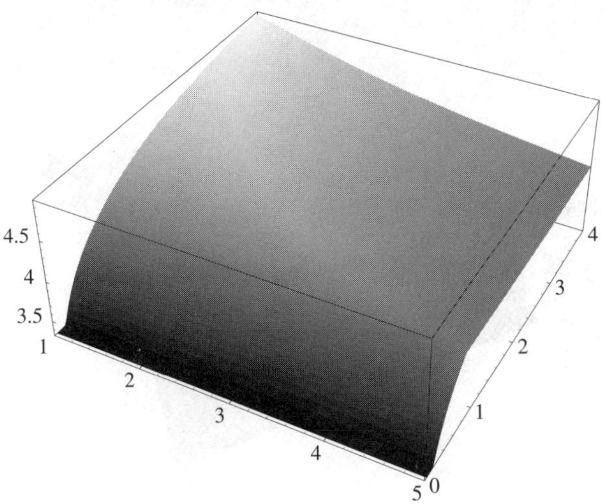

FIG. 8.1.

FIG. 8.2.

$I_1 = \{M_1\}$ and $I_2 = \{M_2, M_3\}$. Both these sets are needed to provide a service to every market and so by complementary slackness $\tau_1 = \tau_2 = 0$.

UNRESTRICTED REGIME: Just as in Example 8.3.2, (8.6) gives $\lambda_1 = a$ for I_1, but for I_2 it gives $\lambda_2 + \lambda_3 = a$. Because both M_2 and M_3 are members of precisely the same I_k (in this case I_2 only), then the prices in the two markets are equal and hence the corresponding λ_i are also equal, that is $\lambda_2 = \lambda_3 = a/2$.

RESTRICTED REGIME: We now have $\lambda_1 = \mu + a$ and $\lambda_2 = \lambda_3 = (\mu + a)/2$, with $y_1 = x_1 = 1/(nP_1^\epsilon)$ and $y_2 = x_2 = x_3 = 1/(nP_2^\epsilon)$. The condition that $y_1 + y_2 = r$ in the restricted regime may be used to determine the values of λ_1, λ_2 and λ_3 (numerically) and also indicates that the boundary between the two regimes occurs when

$$(1 + \lambda_1)^{-\epsilon} + (1 + \lambda_2)^{-\epsilon} = nr\gamma_n^\epsilon\,, \tag{8.16}$$

or, equivalently

$$\frac{1}{(1+a)^\epsilon} + \frac{2^\epsilon}{(2+a)^\epsilon} = nr\gamma_n^\epsilon\,. \tag{8.17}$$

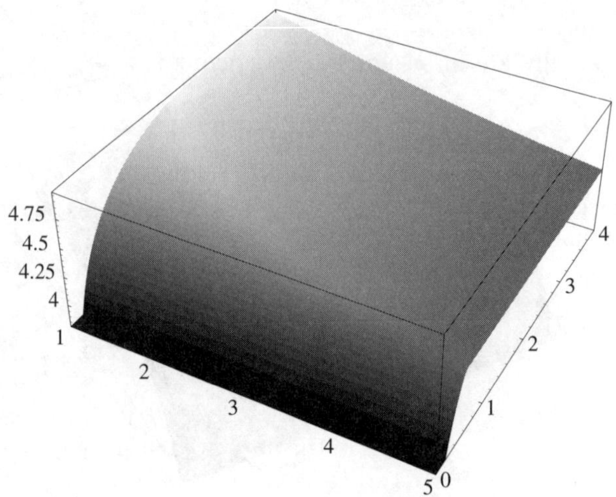

FIG. 8.3.

Figure 8.3 shows the average welfare per market for this interference graph for the same ranges of values of $(1+a)^\epsilon$ and $rn\gamma_n^\epsilon$ as in Fig. 8.1. The new possibility of spectrum reuse between markets M_2 and M_3 means that higher welfare values are possible, especially when the spectrum allocations are low, and also that smaller spectrum allocations are needed in order to reach welfare values on the 'plateau' towards the back of the figure.

8.3.4 Three nodes and one edge

This example corresponds to the second interference graph in Fig. 8.2, where the maximal independent sets are $I_1 = \{M_1, M_3\}$ and $I_2 = \{M_2, M_3\}$. The nonempty intersection of I_1 and I_2 leads to some new possibilities, as explained below. It is still the case that $\tau_1 = \tau_2 = 0$, since both I_1 and I_2 are needed to provide a service in all three markets.

UNRESTRICTED REGIME: Equation (8.6) gives $\lambda_1 + \lambda_3 = \lambda_2 + \lambda_3 = a$ and so $\lambda_1 = \lambda_2$, which in turn means that $x_1 = x_2$ and $P_1 = P_2$. We can rule out the possibility that $\lambda_1 = \lambda_2 = 0$, since then $\lambda_3 = a$ and so $x_3 = y_1 + y_2$ by complementary slackness; but we must have $x_1 \le y_1$ and $x_2 \le y_2$ and so $x_1 + x_2 \le x_3$, which is inconsistent with $\lambda_1, \lambda_2 \le \lambda_3$. Therefore, $\lambda_1, \lambda_2 > 0$ and (by complementary slackness again) $x_1 = y_1 = x_2 = y_2$.

There are now two cases to consider. First, suppose that $\lambda_3 > 0$, so that $x_3 = y_1 + y_2 = 2x_1$. Then $2(1+\lambda_3)^\epsilon = (1+\lambda_1)^\epsilon$, which, together with $\lambda_1 + \lambda_3 = a$ is sufficient to determine the λ_i, namely

$$\lambda_1 = \lambda_2 = \frac{(1+a)2^{1/\epsilon} - 1}{1 + 2^{1/\epsilon}} \quad \text{and} \quad \lambda_3 = \frac{a - (2^{1/\epsilon} - 1)}{1 + 2^{1/\epsilon}}. \tag{8.18}$$

This solution will hold provided that both $(1+a)^\epsilon > 2$ (i.e. $\lambda_3 > 0$) and $y_1+y_2 = x_3 \le r$ (i.e. we really are in an unrestricted regime); this latter condition may be written as

$$nr\gamma^\epsilon \ge \left(\frac{1+2^{1/\epsilon}}{(2+a)}\right)^\epsilon . \tag{8.19}$$

The other case to consider is $\lambda_3 = 0$, that is $\lambda_1 = \lambda_2 = a$. This solution will hold provided that $x_3 \le y_1 + y_2 \le r$. Recalling that $y_1 = y_2 = x_1 = x_2$, these conditions become $x_3 \le 2x_1 \le r$, which may be easily recast as $(1+a)^\epsilon \le 2$ and

$$nr\gamma^\epsilon \ge \frac{2}{(1+a)^\epsilon} . \tag{8.20}$$

RESTRICTED REGIME: Equation (8.6) now gives $\lambda_1+\lambda_3 = \mu+a$ and $\lambda_2+\lambda_3 = \mu+a$, and so again $\lambda_1 = \lambda_2$. Just as in the unrestricted regime, λ_1 and λ_2 cannot be zero and there are similarly two cases to consider regarding the value of λ_3, both having $x_1 = y_1 = x_2 = y_2$.

If $\lambda_3 > 0$ then eqn (8.18) holds for λ_1, λ_2 and λ_3, but with a replaced by $a+\mu$. The defining property of the restricted regime, $y_1 + y_2 = r$, may be recast as $2x_1 = r$ and then gives a further equation for μ. The conditions for this solution to hold, namely $\lambda_3 > 0$ and $\mu > 0$, may then be written as

$$nr\gamma^\epsilon < 1, \quad \text{and} \quad nr\gamma^\epsilon < \left(\frac{1+2^{1/\epsilon}}{(2+a)}\right)^\epsilon . \tag{8.21}$$

If $\lambda_3 = 0$ then $\lambda_1 = \lambda_2 = \mu + a$. As above, we have $x_1 + x_2 = r$ for a restricted regime, which determines μ. This solution holds provided that $\mu \ge 0$ and $x_3 \le y_1 + y_2 = r$, which gives the conditions

$$1 \le nr\gamma_n^\epsilon \le \frac{2}{(1+a)^\epsilon} . \tag{8.22}$$

We now have four regions covering all possible choices for a and r, corresponding to unrestricted or restricted regimes, each with either $\lambda_3 = 0$ or $\lambda_3 > 0$. The significance of the regions in which $\lambda_3 = 0$ is that the favourable interference environment of market M_3 leads to a market price that does not depend on the level of spectrum fees. The spectrum capacity that firms acquire to serve the other markets is reused in M_3. Figure 8.4 shows average welfare in the same way as for the earlier examples. The four regions fit together to give a surface that is in fact very close to that of Example 8.3.3. The welfare is slightly higher here, as one would expect, but visually the difference is noticeable only at very low levels of spectrum allocation.

8.4 Discussion and possible extensions

In this chapter, we have addressed some of the issues involved in constructing a model of radio channel assignment that integrates engineering and economic

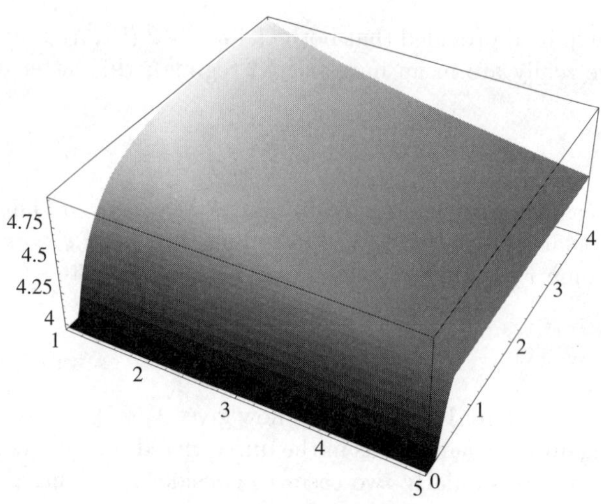

FIG. 8.4.

considerations. It has been shown how regulatory tools such as spectrum fees and spectrum allocations might be tuned to the radio interference environment in order to maximize economic welfare. In any particular example, there are regimes in which the economic and engineering circumstances mean that firms would benefit from an additional allocation of channels (the restricted regime), and regimes in which the spectrum allocation does not restrict output (the unrestricted regime). From the regulatory point of view, it seems sensible to operate near the interface of these regimes.

There are still a number of very natural questions that have not yet been addressed. For example, a generalization to nonidentical firms would shed light on the extent to which firms attach differing values to spectrum according to the scope they have for reusing it in other locations. Within such a model, one would be able to have firms operating in some markets, but not others, and then there are interesting questions of market entry, market exit, and of the migration of firms between markets. A detailed investigation of these questions is likely to require the building of a computational framework in which to automate the analysis and to extend it to much larger examples.

BIBLIOGRAPHY

Aarts, E. and Lenstra, J. (eds) (1997). *Local search in combinatorial optimization*, Wiley, Chichester.

Allen, S. M., Smith, D. H., and Hurley, S. (1999). Lower bounding techniques for frequency assignment, *Discrete Mathematics* **197/198**, 41–52.

Anderson, H. R. and McGeehan, J. P. (1994). Optimizing microcell base station locations using simulated annealing techniques, *Proceedings 44th IEEE Conference on Vehicular Technology*, pp. 858–62.

Anderson, L. G. (1973). A simulation study of some dynamic channel assignment algorithms in a high capacity mobile telecommunications system, *IEEE Transactions on Communications* **21**, 1294–301.

Back, T. (1993). Optimal mutation rates in genetic search, *Proceedings of the 5th International Conference on Genetic Algorithms*, Morgan Kaufmann, pp. 2–8.

—— (1996). *Evolutionary algorithms in theory and practice*, Oxford University Press, New York.

Bell, T. E. (1995). Remote sensing, *IEEE Spectrum* **32**, 25–31.

—— (1996). Main event: Spectrum auctions, *IEEE Spectrum* **33**, 28.

Bellchambers, W. H. (1992). Incorporating flexibility into spectrum allocations, *The Spectrum Mundwrestle—WARC 92 and beyond*, CSIS, pp. 73–4.

Berge, C. (1973). *Graphs and Hypergraphs*, North Holland.

Berger, M. (1995). Neural channel assignment—the fast way, *IEEE International Conference on Neural Networks, Australia*, pp. 1557–60.

Bernstein, M., Sloane, N. J. A., and Wright, P. E. (1997). On sublattices of the hexagonal lattice, *Discrete Mathematics* **170**, 29–39.

Box, F. (1978). A heuristic technique for assigning frequencies to mobile radio nets, *IEEE Transactions on Vehicular Technology* **27**, 57–64.

Boyce, J. F., Dimitropoulos, C. H. D., vom Scheidt, G., and Taylor, J. G. (1995). GENET and tabu search for combinatorial optimization problems, *World Congress on Neural Networks*, Washington DC, USA.

Calégari, P., Guidec, F., and Kuonen, P. (1997*a*). Urban radio network planning for mobile phones, *EPFL Supercomputing Review* **9**, 4–10.

—— —— —— Wagner, D. (1997*b*). Genetic approach to radio network optimizations for mobile systems, *Proceedings of the 47th Vehicular Technology Conference, Phoenix, USA*, pp. 755–9.

Carraghan, R. and Pardalos, P. M. (1990). An exact algorithm for the maximum clique problem, *Operations Research Letters* **9**, 375–82.

Castelino, D. and Stephens, N. (1999). A surrogate constraint tabu thresholding implementation for the frequency assignment problem, *Annals of Operations Research* **86**, 259–70.

Cave, M. (2002). Review of radio spectrum management, independent review conducted for the Department of Trade and Industry.

Chan, P., Palaniswami, M., and Everitt, D. (1991). Dynamic channel assignment for cellular mobile radio sytem using feedforward neural networks, *Proceedings of the International Joint Conference on Neural Networks, Vol. 2*, Singapore, pp. 1242–7.

——————(1994). Neural network-based dynamic channel assignment for cellular mobile communication systems, *IEEE Transactions on Vehicular Technology* **43**, 279–88.

Clark, B. N., Colbourn, C. J., and Johnson, D. S. (1990). Unit disk graphs, *Discrete Mathematics* **86**, 165–77.

Cohen, D. A., Gyssens, M., and Jeavons, P. G. (1996). Derivation of constraints and database relations, *Lecture Notes in Computer Science* **1118**, 134–48.

Cooper, M. C. (1989). An optimal k-consistency algorithm, *Artificial Intelligence* **41**, 89–95.

Costa, D. (1993). On the use of some known methods for T-colourings of graphs, *Annals of Operations Research* **41**, 343–58.

——Hertz, A. (1997). Ants can colour graphs, *Journal of the Operational Research Society* **48**, 295–305.

————Dubois, O. (1995). Embedding a sequential procedure within an evolutionary algorithm for coloring problems in graphs, *Journal of Heuristics* **1**, 105–28.

Crisan, C. and Muhlenbein, H. (1998). The frequency assignment problem: a look at the performance of evolutionary search, *Lecture Notes in Computer Science* **1363**, 263–74.

Crompton, W., Hurley, S., and Stephens, N. (1993). Frequency assignment using a parallel genetic algorithm, *Proceedings of the IEE/IEEE Natural Algorithms in Signal Processing Workshop*, Vol. 2, IEE, pp. 26/1–26/8.

Daudé, H., Flajolet, P., and Vallée, B. (1997). An average-case analysis of the gaussian algorithm for lattice reduction, *Combinatorics, Probability and Computing* **6**, 397–433.

Dechter, R. (1992). From local to global consistency, *Artificial Intelligence* **55**, 87–107.

——Pearl, J. (1988). Network-based heuristics for constraint satisfaction problems, *Artificial Intelligence* **34**, 1–38.

——van Beek, P. (1997). Local and global relational consistency, *Theoretical Computer Science* **173**(1), 283–308.

Del-Re, E., Fantacci, R., and Ronga, L. (1996). A dynamic channel allocation technique based on Hopfield neural networks, *IEEE Transactions on Vehicular Technology* **45**, 26–31.

Dell'Amico, M., Maffioli, F., and Martello, S. (1997). *Annotated bibliographies in combinatorial optimization*, Wiley, Chichester.

Dooling, D. (1994). A quarter century after Apollo landing, *IEEE Spectrum* **31**, 16–29.

Dougan, D. L. (1992). Somewhere over the technology rainbow: spectrum in perspective, *The Spectrum Mudwrestle—WARC 92 and beyond*, CSIS, pp. iii–vi.

Duque-Anton, M., Kunz, D., and Ruber, B. (1993). Channel assignment for cellular radio using simulated annealing, *IEEE Transactions on Vehicular Technology* **42**, 14–21.

Fausett, L. (1994). *Fundamentals of neural networks*, Prentice Hall, Englewood Cliffs.

Fleming, D. J. (1985). State sovereignty and the effective management of a shared universal resource: observations drawn from examining developments in the international regulation of radiocommunications, *Annals of air and space law* **X**, 326–52.

Fogel, D. (1991). *System identification through simulated evolution: a machine learning approach to modelling*, Ginn Press, Needham Heights.

——(1992). *Evolving artificial intelligence*, PhD thesis, University of California, San Diego.

Fogel, L., Owens, A., and Walsh, M. (1966). *Artificial intelligence through simulated evolution*, Wiley, New York.

Freuder, E. C. (1982). A sufficient condition for backtrack-free search, *Journal of the ACM* **29**, 24–32.

——(1985). A sufficient condition for backtrack-bounded search, *Journal of the ACM* **32**, 755–61.

Funabiki, N. and Takefuji, Y. (1992). A neural network parallel algorithm for channel assignment problems in cellular radio networks, *IEEE Transactions on Vehicular Technology* **41**, 430–7.

Gamst, A. (1982). Homogeneous distribution of frequencies in a regular hexagonal cell system, *IEEE Transactions on Vehicular Technology* **31**, 132–44.

——(1986). Some lower bounds for a class of frequency assignment problems, *IEEE Transactions on Vehicular Technology* **35**, 8–14.

Garey, M. R. and Johnson, D. S. (1979). *Computers and intractability: a guide to the theory of NP-completeness*, Freeman, San Francisco.

Gendreau, M., Soriano, P., and Salvail, L. (1993). Solving the maximum clique problem using a tabu search approach, *Annals of Operations Research* **41**, 385–403.

Gerke, S. (2000*a*). Colouring a weighted bipartite graph with a co-site constraint, *Discrete Mathematics* **224**, 115–38.

——(2000*b*). Weighted colouring and channel assignment, D.Phil. thesis, University of Oxford.

Gerke, S. and McDiarmid, C. (2001*a*). Graph imperfection, *Journal of Combinatorial Theory B* **83**, 58–78.

——— (2001*b*). Graph imperfection II, *Journal of Combinatorial Theory B* **83**, 79–101.

Glover, F. (1977). Heuristics for integer programming using surrogate constraints, *Decision Sciences* **8**, 156–66.

—— (1991). Multilevel tabu search and embedded search neighbourhoods for the travelling salesman problem, *Technical report*, Graduate School of Business, University of Colorado at Boulder.

—— (1994). Genetic algorithms and scatter search: unsuspected potentials, *Statistics and Computing* **4**, 131–40.

—— Taillard, E., and de Werra, D. (1993). A user's guide to tabu search, *Annals of Operations Research* **41**, 3–28.

Goldberg, D. (1989). *Genetic algorithms in search, optimization and machine learning*, Addison Wesley.

Gräf, A., Stumpf, M., and Weißenfels, G. (1998). On coloring unit disk graphs, *Algorithmica* **20**, 277–93.

Grötschel, M., Lovász, L., and Schrijver, A. (1993). *Geometric algorithms and combinatorial optimization*, second edn, Springer-Verlag, Berlin.

Gyssens, M., Jeavons, P., and Cohen, D. (1994). Decomposing constraint satisfaction problems using database techniques, *Artificial Intelligence* **66**(1), 57–89.

Hale, W. K. (1980). Frequency assignment: theory and applications, *Proceedings of the IEEE* **68**, 1497–514.

—— (1981). New spectrum management tools, *IEEE International Symposium on Electromagnetic Compatibility*, pp. 47–53.

Hao, J. and Dörne, R. (1996). Study of genetic search for the frequency assignment problem, *Lecture Notes in Computer Science* **1063**, 333–44.

Hardin, G. (1968). The tragedy of commons, *Science* **162**, 1243–8.

Havet, F. (2001). Channel assignment and multicolouring of the induced subgraphs of the triangular lattice, *Discrete Mathematics* **233**, 219–31.

Holland, J. (1975). *Adaption in natural and artificial systems*, University of Michigan Press, Ann Arbor.

Huang, D. C. (1993). Managing the spectrum-win, lose or share, *Technical Report P-93-2*, Centre for Information Policy Research, Harvard University.

Hurley, S. (2000). Automatic base station selection and configuration in mobile networks, *IEEE Vehicular Technology Conference, Fall 2000*, Boston, pp. 2585–92.

—— Smith, D. H., and Thiel, S. U. (1997). FASoft: A system for discrete channel frequency assignment, *Radio Science* **32**, 1921–39.

—— Thiel, S., and Smith, D. (1996). A comparison of local search algorithms for radio link frequency assignment problems, *ACM Symposium on Applied Computing*, pp. 251–7.

ITU (1992). Constitution and convention of the ITU, Article 44, Item 196.

——(1996). Radio regulations.

——(1997). Final acts of the world radiocommunications conference.

Janssen, J. and Kilakos, K. (1996). Polyhedral analysis of channel assignment problems: (I) tours, London School of Economics research report CDAM-96-17.

Jeavons, P. G., Cohen, D. A., and Gyssens, M. (1996). A test for tractability, *Lecture Notes in Computer Science* **1118**, 267–81.

————(1997). Closure properties of constraints, *Journal of the ACM* **44**, 527–48.

Jensen, T. R. and Toft, B. (1995). *Graph Coloring Problems*, Wiley, New York.

Jordan, S. and Schwabe, E. J. (1996). Worst-case performance of cellular channel assignment policies, *ACM Journal on Wireless Networks* **2**, 265–75.

JTAC IRE-RTMA (1952). *Radio spectrum conservation*, McGraw-Hill Book Company.

Kapsalis, A., Rayward-Smith, V., and Smith, G. (1995). Using genetic algorithms to solve the radio link frequency assignment problem, *Proceedings 2nd International Conference on Artificial Neural Networks and Genetic Algorithms*, Springer-Verlag, Arles, France, pp. 37–40.

Kirkpatrick, S., Gelatt, C. D., and Vecchi, M. P. (1983). Optimization by simulated annealing, *Science* **220**, 671–80.

Kunz, D. (1991). Channel assignment for cellular radio using neural networks, *IEEE Transactions on Vehicular Technology* **40**, 188–93.

Ladkin, P. B. and Maddux, R. D. (1994). On binary constraint problems, *Journal of the ACM* **41**, 435–69.

Lagarias, J. C. (1995). Point lattices, in *Handbook of combinatorics*, Graham, R. L., Grötschel, M., and Lovász, L. (ed.) pp. 919–66, Elsevier, Amsterdam.

Lanfear, T. A. (1989). Graph theory and radio frequency assignment, *Technical report*, NATO Headquarters, 1110 Brussels. NATO EMC Analysis project no. 5.

Leese, R. A. (1996). Tiling methods for channel assignment in radio communication networks, *Zeitschrift für Angewandte Mathematik und Mechanik* **76**, 303–6.

——(1997). A unified approach to the assignment of radio channels on a regular hexagonal grid, *IEEE Transactions on Vehicular Technology* **46**, 968–80.

Lochtie, G. D. (1993). Frequency assignment using artificial neural networks, *Proceedings of the 8th International Conference on Antennas and Propagation*, published by IEE, pp. 948–51.

——Mehler, M. J. (1995). Subspace approach to channel assignment in mobile communications networks, *IEE Proceedings on Communications* **142**, 179–85.

Macario, R. C. V. (1997). *Cellular radio—principles and design*, Macmillan Press, London.

MacDonald, V. H. (1979). The cellular concept, *Bell System Technical Journal* **58**, 15–41.

Mackworth, A. K. (1977). Consistency in networks of relations, *Artificial Intelligence* **8**, 99–118.

Mackworth, A. K. (1992). Constraint satisfaction, *in* S. C. Shapiro (ed.), *Encyclopedia of Artificial Intelligence*, Vol. 1, Wiley Interscience, pp. 285–93.

Mathar, R. and Mattfeldt, J. (1993). Channel assignment in cellular radio networks, *IEEE Transactions on Vehicular Technology* **42**, 647–56.

McDiarmid, C. (2000). Frequency-distance constraints with large distances, *Discrete Mathematics* **223**, 227–51.

——(2001). Channel assignment and graph imperfection, in *Perfect Graphs*, Ramírez-Alfonsín, J. L. and Reed, B. A. (ed.) pp. 215–31, Wiley, New York.

——Reed, B. (1999). Colouring proximity graphs in the plane, *Discrete Mathematics* **199**, 123–37.

————(2000). Channel assignment and weighted colouring, *Networks* **36**, 114–7.

Metropolis, N., Rosenbluth, A., Rosenbluth, M., Teller, A., and Teller, E. (1953). Equations of state calculations by fast computing machines, *Journal of Chemical Physics* **21**, 1087–92.

Molina, A., Athanasiadou, G., and Nix, A. (1999). The automatic location of base-stations for optimised cellular coverage: A new combinatorial approach, *IEEE*.

Montanari, U. (1974). Networks of constraints: fundamental properties and applications to picture processing, *Information Sciences* **7**, 95–132.

Narayanan, L. and Shende, S. (2001). Static frequency assignment in cellular networks, *Algorithmica* **29**, 396–409.

Pach, J. and Agarwal, P. K. (1995). *Combinatorial geometry*, Wiley, New York.

Pekny, J. F. and Miller, D. L. (1994). A staged primal-dual algorithm for finding a minimum cost perfect two-matching in an undirected graph, *ORSA Journal on Computing* **6**, 68–81.

Prim, R. C. (1957). Shortest connection networks and some generalizations, *Bell System Technical Journal* **36**, 1389–401.

Raychaudhuri, A. (1985). *Intersection assignments, T-colourings and powers of graphs*, PhD thesis, Rutgers University.

Rechenberg, I. (1973). *Evolutionsstrategie: optimierung technischer systeme nach prinzipies der biologischen evolution*, Frommann-Holzborg, Stuttgart.

Reeves, C. R. (ed.). (1995). *Modern heuristic techniques for combinatorial problems*, Advanced topics in computer science, McGraw-Hill.

Robinson, G. O. (1980). Regulating international airwaves: The 1979 WARC, *Viginia Journal International Law* **21**, 1–54.

Rodosek, R. (1997). Generation and comparison of constraint-based heuristics using the structure of constraints, Ph.D. thesis, Imperial College, University of London.

Rogers, C. A. (1964). *Packing and covering*, Cambridge University Press, Cambridge.

Schaefer, T. J. (1978). The complexity of satisfiability problems, *Proceedings 10th ACM Symposium on Theory of Computing*, ACM Press, New York, pp. 216–26.

Schiex, T., Regin, J.-C., Gaspin, C., and Verfaille, G. (1996). Lazy arc consistency, *Proceedings of 13th National Conference on Artificial Intelligence AAAI96*, AAAI Press/The MIT Press, pp. 216–21.

Schnabel, N., Ubéda, S., and Žerovnik, J. (1999). A note on upper bounds for the span of the frequency planning in cellular networks, Manuscript.

Schneider, R. (1993). *Convex bodies: the Brunn-Minkowski theory*, Vol. 44 of *Encyclopedia of Mathematics and its Applications*, Cambridge University Press.

Schwefel, H.-P. (1977). *Numerische optimierung von computer modellen mittels der evolutionsstrategie*, Vol. 26 of *Interdisciplinary Systems*, Birkhauser, Basel.

—— (1981). *Numerical optimization of computer models*, Wiley, Chichester.

Shepherd, M. A. (1998). *Radio channel assignment*, PhD thesis, University of Oxford.

Sivarajan, K. N., McEliece, R. J., and Ketchum, J. W. (1989). Channel assignment in cellular radio, *Proceedings of the 39th Conference of the IEEE Vehicular Technology Society*, pp. 846–59.

Smith, D. H. and Hurley, S. (1997). Bounds for the frequency assignment problem, *Discrete Mathematics* **167/168**, 571–82.

—— —— Allen, S. M. (2000). A new lower bound for the channel assignment problem, *IEEE Transactions on Vehicular Technology* **49**, 1265–72.

—— —— Thiel, S. U. (1998). Improving heuristics for the frequency assignment problem, *European Journal of Operational Research* **107**, 76–86.

Struzak, R. (1985). Vestigial radiation from industrial, scientific and medical radio frequency equipment, in *Nonlinear and environmental electromagnetics*, Kikuchi, H. (ed.) pp. 223–52.

—— (1992). Microcomputer modeling, analysis and planning in terrestrial television broadcasting, *Telecommunication Journal* **59**, 459–92.

—— (1993). On future information system for management of radio frequency spectrum resource, *Telecommunication Journal* **60**, 429–37.

Tcha, D.-W., Chung, Y.-J., and Choi, T.-J. (1997). A new lower bound for the frequency assignment problem, *IEEE/ACM Transactions on Networking* **5**, 34–9.

Tsang, E. (1993). *Foundations of constraint satisfaction*, Academic Press, London.

van den Heuvel, J. (1998). Radio channel assignment on 2-dimensional lattices, CDAM Research Reports, **LSE-CDAM-98-05**, London School of Economics, London.

—— Leese, R. A., and Shepherd, M. A. (1998). Graph-labeling and radio channel assignment, *Journal of Graph Theory* **29**, 263–83.

van Hentenryck, P., Deville, Y., and Teng, C.-M. (1992). A generic arc-consistency algorithm and its specializations, *Artificial Intelligence* **57**, 291–321.

Volgenant, A. (1990). Symmetric traveling salesman problems, *European Journal of Operational Research* **49**, 153–4.

Volgenant, T. and Jonker, R. (1982). A branch and bound algorithm for the symmetric traveling salesman problem based on the 1-tree relaxation, *European Journal of Operational Research* **9**, 83–9.

Wang, W. and Rushforth, C. (1996). An adaptive local search algorithm for the channel assignment problem (CAP). *IEEE Transactions on Vehicular Technology* **45**, 459–66.

Webbing, D. W. (1977). The value of the frequency spectrum allocated to specific uses, *IEEE Transactions* **EMC-19**, 343–51.

Zimmermann, H. (1995). Towards the unrestricted use of telecommunications facilities before, during and after an emergency, *The vital role of telecommunications in disaster relief and mitigation.* UN Department of Humanitarian Affairs.

Zoellner, J. A. and Beall, C. L. (1977). A breakthrough in spectrum conserving frequency assignment technology, *IEEE Transactions on Electromagnetic Compatibility* **19**, 313–19.

Zorpette, G. (1995). Radio astronomy new windows on the Universe, *IEEE Spectrum* **32**, 18–25.

INDEX